genetics

a beginner's guide

ONEWORLD BEGINNER'S GUIDES combine an original, inventive, and engaging approach with expert analysis on subjects ranging from art and history to religion and politics, and everything in-between. Innovative and affordable, books in the series are perfect for anyone curious about the way the world works and the big ideas of our time.

aesthetics
africa
american politics
anarchism
animal behaviour
anthropology
anti-capitalism
aquinas
archaeology
art
artificial intelligence
the baha'i faith
the beat generation
the bible
biodiversity
bioterror & biowarfare
the brain
british politics
the Buddha
cancer
censorship
christianity
civil liberties
classical music
climate change
cloning
the cold war
conservation
crimes against humanity
criminal psychology
critical thinking
the crusades
daoism
democracy
descartes
dewey
dyslexia
energy
engineering
the english civil wars
the enlightenment
epistemology
ethics
the european union
evolution
evolutionary psychology
existentialism
fair trade
feminism
forensic science
french literature
the french revolution
genetics
global terrorism
hinduism
the history of medicine
history of science
homer
humanism
huxley
international relations
iran
islamic philosophy
the islamic veil
journalism
judaism
lacan
life in the universe
literary theory
machiavelli
mafia & organized crime
magic
marx
medieval philosophy
the middle east
modern slavery
NATO
the new testament
nietzsche
nineteenth-century art
the northern ireland conflict
nutrition
oil
opera
the palestine–israeli conflict
particle physics
paul
philosophy
philosophy of mind
philosophy of religion
philosophy of science
planet earth
postmodernism
psychology
quantum physics
the qur'an
racism
reductionism
religion
renaissance art
the roman empire
the russian revolution
shakespeare
the small arms trade
sufism
the torah
the united nations
volcanoes
world war II

genetics

a beginner's guide

ONEWORLD

genetics: a beginner's guide

Oneworld Publications
10 Bloomsbury Street
London WC1B 3SR
England

Reprinted 2004, 2011, 2015

A CIP record for this title is available
from the British Library

ISBN 978-1-85168-304-8

Cover design by Two Associates
Typeset by Saxon Graphics Ltd, Derby, UK
Printed and bound in Great Britain by Clays Ltd, St Ives plc

contents

preface

Human life today has been enormously impacted by advances in the science of genetics and related work on the physiology of human reproduction. This book is intended for the average citizen who wants to learn more about the basic science and the critical issues it has raised. The reader we kept in mind as we wrote is a reasonably well-educated person who has probably forgotten the bits of genetics he or she may have studied in school. The book presents the basic concepts of genetics and develops some background for understanding current controversies over genetic manipulation of food and even of humans themselves.

We moderns tend to be very shortsighted. We tend to think that interest in genetics, and knowledge of it, are very recent phenomena, and that problems relating to heredity and reproduction have only arisen in the last few decades, with the development of molecular genetics. One purpose of this book is to dispel that belief. We have reached far back into human history, to ancient myths and writings, to traces of art left by our long-forgotten ancestors, which provide some insight into what they were thinking and feeling. In fact, of course, concerns about heredity and reproduction go back to the beginnings of our species. Reproduction is the prime concern of every species, even if its members lack the consciousness to be aware of it. We cannot know what our habiline or erectus ancestors may have thought millions of years ago as they started to become aware of themselves and of the problems of existence; but at some remote time hominids had to realize that they depended upon the continuous birth of new people, and they must have begun to wonder how reproduction occurs and to be concerned about the birth of healthy babies who looked the way human babies are supposed to look. As we show, concerns about heredity and attempts to control it developed strongly when domestication of plants and animals began.

We devote some words to insights revealed by art and literature. We do not consider these matters trivial. They are integral to human knowledge, and we have tried to include them wherever it seemed appropriate. We have thus tried to write a book that will appeal broadly to people who want to understand science within the wider context of human culture. Interspersed among the arts and history, and as an adjunct to the straightforward presentation of science, we have also tried to present a realistic conception of how modern science is done. It is still one of the most exciting of human activities, and its story deserves to be told. At the same time, people must understand its logic and its boundaries, so science can be seen in proper perspective as a cultural phenomenon.

We have tried to present a balanced view of the controversies engendered by modern genetics. In fact, we have not always agreed with one another in writing this book, and we have achieved some of this balance in trying to work out a text that we could all accept. If biases remain, they are the biases of a liberal humanism which are, we think, justified by biology. We are not cheerleaders for science, because we recognize the inherent dangers in almost every scientific innovation; but neither are we Luddites. When recombinant-DNA technology was invented, many respected scientists warned of possible disasters. In retrospect, it was important for scientists to foresee the dangers and guard against them; but when reasonable controls were developed, mandated by the fears of disaster, it became clear that humanity might well receive the benefits of the technology without its dangers. It seems most reasonable now to continue with new developments in the same vein. However, every technology engenders serious social and ethical questions, which rational and informed people must debate. We have tried to at least bring these questions to the reader's attention and to present some of the major viewpoints that have already been expressed.

We also believe in humanity as a species and in the essential biological equality of all peoples. In an era when people worldwide are being harrassed for having the wrong color of skin or speaking in the wrong language, when neo-Nazis seek to spread their venom in Iowa and Idaho, it is important for scientists to recognize the "moral un-neutrality of science," as the scientist and novelist C. P. Snow once put it, and to relate clearly what biology says. As we read it, it shows enormous genetic variance within and between groups, but all falling within a common range that points to no innate inferiority or superiority of any group. We at least want to speak a moral truth to readers who may be confused about these matters.

Burton S. Guttman, Anthony J. F. Griffiths
David T. Suzuki, Tara Cullis

July, 2002

chapter one

genetics: past, present, and future

"Why does Jimmy have red hair like his mommy when his daddy has brown hair?"

"Why don't people have baby puppies?"

"Can a horse marry a cow and have babies?"

"Why is Mary so tall when her parents are so short?"

A child's questions, so free of preconceptions, often penetrate to the very heart of life's most profound mysteries. In fact, such innocent questions have occupied philosophers and scientists since antiquity. The answers to those questions have become embedded in our myths, superstitions, and the conventional wisdom called common sense.

We all take for granted that living organisms perpetuate their species – that generation after generation, cows always have calves, carrot seeds grow into carrot plants, and women give birth to baby humans. For the authors of the Bible, such faithful reproduction of each species was sufficiently impressive to merit mention as a divine injunction in the book of Genesis (1:11, 21):

> And God said, "Let the earth bring forth vegetation, plants yielding seed, and fruit trees bearing fruit in which is their seed, *each according to its own kind,* upon the earth."

> So God created the great sea monsters and every living creature that moves, with which the waters swarm, *according to their kinds* and every winged bird *according to its kind.*

Yet the individuals within a species vary tremendously in form and appearance. Just look at the diversity among people on a city street.

As people reproduce, they create children who not only look human, but look very much like their parents. We carry within us not simply an injunction to reproduce "after our kind," but to reproduce specific features of height, weight, skin color, eyes, hair, and so on. People always took this fact for granted while searching for an explanation, an explanation that for a long time eluded them. And so they fell back on explanations rooted in myths and superstition.

The spectrum of human variation is so broad that one might suppose a woman could occasionally give birth to something that does not look human. Indeed, occasional babies with severe abnormalities *are* born, but they are so rare that popular imagination has often turned them into fantastic mythological creatures. Humans almost always faithfully beget ordinary humans, yet with such great variation that almost every newborn child is unique. How can there be both rigid constancy and boundless difference? Only our insights into the basis of heredity have resolved this apparent biological paradox. The discipline that studies heredity and searches for the principles governing inheritance is called *genetics*.

The modern science of genetics began in 1900, when the fundamental laws determining the transmission of hereditary traits from one generation to the next were discovered. These laws, which apply to all plants and animals as well as many microorganisms, demonstrate the fundamental similarities among life forms. Furthermore, these insights give people enormous power to manipulate living organisms. Practical geneticists have successfully bred high-yielding strains of domestic animals, plants, and antibiotic-producing fungi, and exotic varieties of flowers and goldfish. As we have come to understand the molecular basis of life, our ability to engineer the biological makeup of organisms has passed from science fiction to actual science. News stories almost daily herald the age of genetic engineering.

Applying hereditary principles to humans, we have come to understand the basis for many inherited diseases as well as physical and behavioral traits. These insights penetrate to the very essence of human nature; and just as our basic knowledge in endocrinology, physiology, and embryology has been applied to understanding people, so will our understanding of genetics. Yet the same knowledge has already raised profound moral and ethical issues. In what situations, for instance, will prospective parents consider an abortion – a severe physical or mental defect, a cleft lip, or even an unfavoured gender? When does a developing embryo become human? (Or is this a meaningful question?) In the light of the stupendous power of the first atomic bomb, Aldous Huxley recognized the far greater potential of human engineering in his 1947 foreword to *Brave New World*:

> The release of atomic energy marks a great revolution in human history but not the final and most searching revolution… The really revolutionary revolution is to be achieved not in the external world, but in the souls and flesh of human beings.

That portentous forecast is now being realized. As we enter a new era, we would do well to reflect on the historical and social context surrounding this new technology.

the search for order and meaning

The microbial geneticist François Jacob observed that "It is a requirement of the human brain to put order in the universe." Every infant begins to perceive the world without any framework with which to make sense of its experiences. But quickly, through language, the child learns to fit its observations into society's scheme of things, whether the child is a Stone Age Yanomami of the Brazil–Venezuela border, a teenager in a wealthy white family in Dallas, or a black child in Harlem in New York City. Without such a framework, life would be meaningless, and as humans evolved language and increased their ability to conceptualize, they always tried to create order in their world.

Primitive humans, perceiving the mysterious world around them, tried to connect themselves to their society and to the universe by devising imaginative explanations for how they came to exist, how they are related to the animals and plants around them, how children are produced, and why they resemble or differ from their parents. Rather than take their existence and characteristics for granted, they always sought some explanation, however far-fetched, for these riddles. The unknown is unpredictable and full of terrors; primitive people sought to combat these terrors, to replace the sensation of chaos with one of order. Because the sensation of meaning comes from interrelating bits of information, preliterate humans *totalize*, as Claude Lévi-Strauss puts it, by coordinating all facets of their experience into a unified body of knowledge. They devise complex, ingenious classification schemes that explain everything in their world by interrelating them as much as possible through analogies and perceived similarities. As people observed nature carefully and puzzled over its mechanisms, the explanations they devised for natural phenomena became embodied in all-encompassing frameworks of elaborate myths, legends, and religious ideas, which provided answers to questions they could not have answered in any other way. These combinations of observed fact and often inspired imagination were the forerunners of science. They were hypotheses and theories which in their time were accepted as truth, either literal or figurative. They were also the forerunners of literature and art.

We tend to think of myths as rather silly old stories about the adventures and misadventures of gods, warriors, and demons, invented by primitive people to explain a world they could not understand in our modern, scientific sense. But it is a mistake to dismiss these stories as trivial and old-fashioned, with no more important meaning for humanity. Scholars such as Joseph Campbell and Claude Lévi-Strauss have shown that common themes in myths from many diverse cultures speak to us about the universal concerns of all people and about ways of thought that all humans share. As we move increasingly toward a unified world – a global village, as it has been called – it is important to see how much basic human nature we all share. The systematic study of mythology reveals important points about the human psyche, about universal human motivations, fears, and thought patterns. Though we cannot explore all these matters here, as students of human biology we must factor them into our overall conception of how humans function.

Furthermore, Mark Schorer has proposed a broader idea of mythology: "a large controlling image that gives philosophic meaning to the facts of ordinary life; that is, that has organizing value for experience… All real convictions involve a mythology."[1] In this sense, all of science is one kind of mythology. Those of us who are engaged in scientific work believe strongly in the value of our enterprise; we believe it is generally good to acquire greater knowledge and a deeper understanding of the natural world, and the knowledge we have gained informs our lives and colors our view of the world. We believe the natural world consists of real entities, entities linked to one another through a complex causal structure whose details we seek to understand. Furthermore, the pursuit of scientific knowledge of the world gives meaning to our daily lives. These convictions are neither true nor false. They are not contentions about the world that we try to prove but are, rather, guidelines for conducting our work or value judgements about what activities are useful and satisfying.

Schorer's words emphasize mythology as a schema that helps us understand the particular events of life. Science deals with the general, the universal. The law of gravitation says that objects fall toward one another in accordance with their masses and the distance between them, and experiments show that objects fall toward the Earth with an acceleration of 9.8 meters per second per second. So if a flower pot falls from a window we can predict when it will reach the sidewalk and what force it will exert. But science says nothing more about individual events; it does not explain why a flower pot fell out of a window just as you passed by and struck you on the head. Yet people ask questions about these events: "Why me?" "Of all the gin joints in all the towns in all the world she walks into mine," laments Rick in *Casablanca*, and we all tend to look for meanings in simple events. Science does not supply meanings. So – unless we are content to simply ascribe events to chance and look no deeper – we are inclined to look elsewhere, generally into some other kind of mythology.

Today our knowledge is fragmented into isolated compartments of science, art, business, ritual, religion, and mythology. Instead of one totalized, unified system, we have many independent systems which few minds are capable of interrelating. It is not surprising, then, that the sensation of meaninglessness constantly threatens to return.

the modern image of science

Genetics is one important aspect of modern biology, and to understand it, we need to put science in general into context. Science is a human activity and a major feature of all human cultures. Gathering and organizing knowledge is perhaps the most characteristic trait of humans, and one of our goals in this book is to show more clearly just how science is done and what a powerful and exciting activity it can be. But it is also important to see that human culture entails other activities that are not – and never can be – science.

Although science is fundamentally about understanding how the natural world operates, it has developed with the ideal that knowledge should benefit humanity, and science has been heavily weighted toward controlling nature to improve human life. By the time the first books of the Old Testament were being written, human historical experience of nature was codified, in the words of Genesis again, into two injunctions with far-reaching implications for civilization:

> Be fruitful and multiply, and fill the earth and subdue it; and have dominion over the fish of the sea and over the birds of the air and over every living thing that moves upon the earth. (Gen. 1:28)

Clearly, the original seeds of the population explosion are contained in the first command. In the second, the role of humans in the Judeo-Christian tradition is established: we are to be masters over nature.

As we shall see in Chapter 2, human interest in genetics can be traced to the beginnings of agriculture in Neolithic times around 9000–7000 BC, when people began to domesticate plants and animals. They soon realized that they could improve their crops and herds through selective breeding. Recognizing that plants and animals could be improved so dramatically, thoughtful people began to consider how humans could improve themselves as well. In ancient Greece, Plutarch records that Lycurgus, the founder of Sparta, set up boards to determine which couples were likely to reproduce offspring embodying Spartan ideals. Babies judged to fall below standard were left to die at the foot of Mt Taegetus. Even in intellectual, aesthetic Athens, Socrates remarks that if it is important to breed hunting dogs and birds with care, then surely the state should be just as careful in the breeding of its maidens and men. Although these ideas never reached practice on a large scale, their continuing debate through the millennia attest to human aspirations to perfection, and we shall address them again in Chapter 15.

The road to perfecting humans and nature clearly required knowledge. In the dictum that "knowledge is power," the seventeenth-century English philosopher Francis Bacon expressed the realization that scientific knowledge, when harnessed through technology, could be a powerful force for human progress. Progress, to Bacon, was measured by the degree to which the biblical injunction to dominate nature was met. Although Bacon's influence is probably overemphasized in modern accounts, his emphasis on the experimental method and his ideal of progress had considerable influence on scientists of the Royal Society (founded in 1662). Baconian ideas became an important inspiration for the Scientific Revolution and an important part of the philosophical orientation of science.

Today the most explosive force in society is the technological application of science. People have reacted to modern science in two essentially opposite ways, according to their temperament: either to embrace science completely in an almost religious way or to reject it out of fear. In Rodgers and Hammerstein's play *The King and I*, the King of Siam – who is trying to modernize his country – approves of many ideas as being "scientific" and reproves Anna, the English governess, for ideas that are "not scientific." Advertisers try to sell their products by using our tendency to think of anything "scientific" as excellent and admirable. And there are scientists and cheerleaders for science who uphold this attitude, called *scientism*, and who believe that science can and should be all-pervasive, providing the answers to everything. They would try to develop equations for emotion or beauty, to produce art with intelligent computers, to explain the beauty of a sunset simply as the result of neural circuitry. They might advocate that we control eccentricity and nonconformity with drugs, electrodes, and selective breeding.

This is foolishness. Human activities are not solely directed toward answering questions and gaining knowledge. We can represent the various facets of human activity thus:

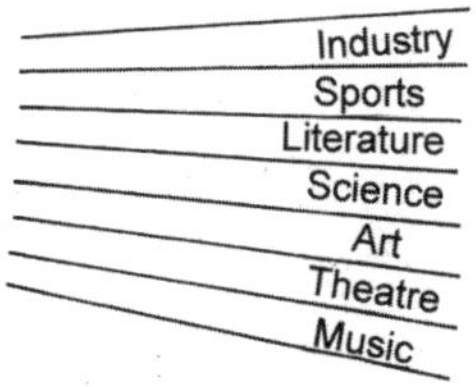

Industry
Sports
Literature
Science
Art
Theatre
Music

The wedge labeled "science" is one area of human activity among many. Many critics of science cry loudly that we must recognize what the limitations to science are; however, the drawing shows that each activity has *boundaries* but no limits. Consider, for a moment, how science compares with the domain of the arts and humanities. Neither one, we think, has limitations. We see no limits to our ability to describe, understand, and find regularities in the universe – the concern of science. We see no limits to the ability of creative humans to find new forms of art or music or literature and new things to do with these forms. The two areas are closely related. The arts may be nourished by contributions from science, such as new technologies and new views of the world, and much of science is a kind of art, done with strong aesthetic ideals in mind. The two are sometimes virtually one, as in the exploration of human form by da Vinci or of the nature of space and light by Cezanne and the Impressionist painters. But fundamentally they are two different human activities, done for different purposes. Neither could supplant the other. Both art and science can enrich daily human activity – the simple business of living – but neither one could supplant it. Even if we fully understood precisely what happens in our nervous systems when we watch a sunset or listen to music or experience love, and no matter how well those experiences are imitated or supplemented by artists and writers, we would still want to go on watching sunsets and listening to music and being in love.

In the wedge labeled "morality" or "ethics," people ask a particularly difficult kind of question – not "What do people do and why?" but "What *should* people do?" The contacts between this wedge and that of science are complex; exploring them will be a major focus of this book. We will try to show how questions of morality must govern the activities of science and how the knowledge gained through science impacts on the moral questions and even creates new moral problems. We cannot give answers to these problems here, but we can at least sort them out a little and try to show where the difficulties and the interesting questions lie.

Thus we reject scientism. We try to see science in its proper place as one human endeavor among many. What, then, of the opposite reaction to science: to fear it and reject it? Science and technology have changed the world radically. People born over sixty years ago came into a society that knew nothing of jet planes, DDT, plastics, television, nuclear bombs, transistors, lasers, computers, satellites, birth-control pills, heart transplants, polio vaccines, or prenatal diagnosis. This flood of information and technology has been overwhelming, and people may feel that we have no more ability to cope with this force than would Neanderthal men given guns. In a sense the Baconian ideal has brought us full circle to a world as terrifying and chaotic as it was at the dawn of human consciousness. Science seems to be taking over and transforming all of human life, yet somehow robbing life of its richness and still failing to provide answers to important moral questions. Faced with such a situation in the past, people have turned to gods, to priests, to oracles who promised to provide answers. Today, however, the people we turn to for explanations and control of the vagaries of nature are no longer gods, but scientists – fallible, mortal beings. Thus it is ironic – but understandable – that in a world in which science is a dominating force so many people have reacted against it by turning to a variety of superstitions and quasi-religions. In a world informed by physics and mineralogy, they believe in astrology and the mystical powers of crystals. In a world informed by physiology and molecular biology, they believe in a host of supposed healing practices from iridology to reflexology.

Ironically, the very science that Bacon envisioned as the means to the full flowering of God's works became the greatest threat to organized religion. Copernicus, Kepler, and Galileo paved the way for Newton while demolishing the church's position that the earth is the center of the universe. Geologists pushed the planet's age further and further beyond Bishop James Usher's date for the Creation (23 October 4004 BC) and Charles Darwin's theory of evolution undermined the biblical story of creation. The church, armed with its inspired writings, chose to fight for its version of truth against the experimental and observational evidence of scientists. When religion lost in this arena, its moral authority seemed to decay as well, leaving a moral vacuum within which scientists continue to apply their knowledge to control and subdue nature. Tragically, organized religion had entered an arena for which it was never intended, just as science was never designed to provide answers to important questions of ethics.

A common criticism of science is that it is done in a cultural void, with no concern for its social repercussions. One typical criticism runs like this:

> Modern science has been singularly devoid of any serious concern with fundamental questions – for example, those involving the

> relations between ends and means. Its overriding instrumentalism has been expressed in its desire to control and dominate nature, almost as an end unto itself.[2]

This criticism has often been legitimate, at least when directed at individuals who have sought to pursue their research programs quite single-mindedly. We shall have occasion in this book to look at such cases and consider their implications, but to put this criticism in perspective we need to distinguish types of scientists and places where research is done. It is estimated that ninety-five percent of all scientists who have ever lived are still actively carrying out research and publishing. Relatively few scientists are now academics, employed by colleges, universities, and institutes dedicated to basic research; yet this is where the bulk of research is done that produces the great fundamental advances in our knowledge of the world. Over half of all scientists and engineers now work full time or carry out research for the military, and most of the remainder work for private industry – including, now, industries devoted to genetic engineering. Thus *power* and *profit* have become primary motivations underlying the application of science, leaving the well-being of the general public and the long-term interests of society and the environment as incidental priorities.

With few exceptions, basic science as a whole has a track record of keeping the cultural context in view. Bacon's own ideal of science emphasized that it must always be done with compassion, with the improvement of human society in mind. And modern science certainly passed a watershed in social awareness with the development of atomic weapons during the Second World War; as J. Robert Oppenheimer put it, "The physicists have known sin." As we shall see, when recombinant-DNA methods were invented – the principal methods that underlie modern genetic research – the scientific community itself was quick to recognize the social implications and potential dangers and to police itself, even if some individuals did not agree to the necessity.

the prospects of modern genetics

In this socio-historical context, we can see why there is such a fascination with genetics and why its social implications are so great. The molecular basis of inheritance and its genetic code have been deciphered; synthetic genes have been created, viruses reproduced in test tubes, identical twins of frogs and sheep created from mature organisms, human eggs fertilized in test tubes, babies inspected inside women's wombs, many hereditary defects cured, and rat–mouse chimeras created.

People are intrigued and fascinated by all this work because we seem to be approaching not only a complete understanding of life itself, but also the potential to alter its properties, to direct evolution. Although even Neolithic farmers directed the evolution of plants and animals by selective breeding, science now offers the possibilities of making novel organisms to serve human needs: plants that synthesize their own fertilizer from the air, bacteria that make human proteins, others that degrade pollutants or make protein from oil, or viruses that carry human genes. As in all fields of science, our knowledge can be used for the benefit or the detriment of humanity. The ability to correct hereditary disease provides a matching ability to induce disease, while the ability to detect and prevent birth defects poses questions: what is to be considered defective, and who decides? The grotesque application of genetic concepts in the race-purification programs of Nazi Germany still commands support in racist and fascist groups throughout the world. And when we manipulate the genetic makeup of organisms, the properties of the creatures produced will not always be predictable.

The dangers inherent in this technology are movingly highlighted for many thoughtful people today by Mary Shelley's classic novel *Frankenstein, or the Modern Prometheus*, which, as Theodore Roszak observes, stands as an allegory for modern science:

> Where did the doctor's great project go wrong? Not in his intentions, which were beneficent, but in the dangerous haste and egotistic myopia with which he pursued his goal. It is both a beautiful and a terrible aspect of our humanity, this capacity to be carried away by an idea. For all the best reasons, Victor Frankenstein wished to create a new and improved human type. What he knew was the secret of his creature's physical assemblage; he knew how to manipulate the material parts of nature to achieve an astonishing result. What he did not know was the secret of personality in nature. Yet he raced ahead, eager to play God, without knowing God's most divine mystery. So he created something that was soulless. And when that monstrous thing appealed to him for the one gift that might redeem it from monstrosity, Frankenstein discovered to his horror that, for all his genius, it was not within him to

> provide that gift. Nothing in his science could comprehend it. The gift was love. The doctor knew everything there was to know about this creature – except how to love it as a person.[3]

Dr. Faustus, Dr. Frankenstein, Dr. Moreau, Dr. Jekyll, Dr. Cyclops, Dr. Caligari, Dr. Strangelove – the scientist who does not face up to the warning in this persistent folklore of mad doctors is the worst enemy of science. In these images of our popular culture resides a legitimate public fear of the scientist's stripped-down, depersonalized conception of knowledge – a fear that scientists, well-intentioned and decent men and women all, will go on being titans who create monsters.

Science cannot be carried out in isolation from society; its priorities and questions are molded by, and in turn affect, the culture of which it is a part. We believe that only through an educated citizenry will profit and power be tempered by the concerns of the public. Our purpose in this book is to set the study of heredity in a social and historical context so its current position and implications for the future of humanity will be better understood.

Now let us turn to the first glimmerings of human consciousness to see how early people attempted to answer their questions about origins, reproduction, and offspring.

chapter two

from myth to modern science

The context of genetics stretches far back into the human past, for the science did not suddenly spring from nowhere at the turn of the twentieth century. Humans probably wondered about inheritance from the dawn of consciousness. Indeed, civilization was made possible largely because people realized that animals and plants could be selectively bred.

For a long time, primitive people survived as our primate ancestors and relatives had done, by hunting and by gathering whatever food they could find. But the most consistent feature of human evolution has been the development of a massive and complex brain that can recognize order in nature. This brain allows us to remember, to learn from others, to avoid repeating past mistakes, and to improve on an advantageous discovery. It was only about ten thousand years ago that some Neolithic people – probably women, who tended the camps while their men were off hunting – discovered the basic idea of domesticating and cultivating plants. At many sites, such as Jericho in the Jordan valley and the annually flooded plains of Egypt, humans realized that seeds planted in soil with sufficient water would grow into useful plants. With such a reliable source of food, they would no longer have to be constantly moving, gathering, and hunting. And so people changed from nomads to settled cultivators.

The settlers selected and nurtured better plants, at first unconsciously. Scavenging edible gourds and fruits from afar, they scattered the seeds, which sprouted into plants that could be domesticated. Wild animals such as dogs, goats, cattle, and sheep were attracted to the crops and quickly discovered scraps and leftovers. Some were captured and kept in pens to provide a dependable supply of meat, hide, and muscle power. As the settlers cleared more land, they left fruit and nut trees standing to become the first orchards. *Homo sapiens* had become a farmer.

People probably attempted sporadic experiments in agriculture time and again around the globe, with many ending in failure. Agriculture finally developed in two broad areas: southern Asia from Mesopotamia to China between about 9000 and 7000 BC, and America from Mexico to Peru around 5000–2000 BC. The New World contribution should not be underestimated, for about sixty percent of the plants currently grown in agriculture were unknown in Europe before the voyages of Columbus. The list of livestock and domestic plants used today represents only a small part of those that have ever been tried.

Ancient humans realized, moreover, that plants and animals reproduce "each according to its kind," so seeds from plants with bigger fruits tend to produce plants bearing bigger fruits, and the offspring of sheep with finer fur will have similar wool. Once people understood the principle of like begetting like, they could harness nature for human benefit.

The effects of agriculture are staggering. This intellectual breakthrough revolutionized human evolution, for now the explosive force of *cultural* evolution, not *biological* evolution, became the shaping power in human history. The domestication of plants and animals provided a community with stability, since people no longer depended on foraging and hunting. With a burgeoning population and the cultivation of large fields, villagers could concentrate their skills on special functions to supply the increasingly diverse needs of the community. They needed defenses against the depredations of nomads; they needed implements to cultivate, water, harvest, and store crops, and to scavenge raw materials for building from increasingly greater distances. The division of labor released more time for the reflection, imagination, and invention that spur the development of culture. This leisure opened up entirely new avenues for experimentation and change: pottery, weaving, smelting, and the wheel, each accelerating human control over the environment and human destiny. Civilization, then, became possible when nomadic hunters and foragers were transformed into farmers who could apply simple rules of plant and animal breeding.

primitive interest in heredity

Looking into the past for early evidence of interest in heredity, we find that even Paleolithic humans had grasped the basics of reproduction. Depictions of mating and reproducing animals and humans on the walls of caverns, for example, performed a dual role. They encouraged the reproduction of humans and of hunted species through sympathetic magic, while educating the young in the appearance, life cycles, and habits of their own race and of the animals on which their survival would depend. These depictions suggest that the ancients had already started to understand some genetic principles when they carved, painted, and wrote their legends and mythologies. When they developed the myth cycles that have come down to us, people already knew they could increase desired characteristics of animals and plants by carefully choosing parent stocks. So these myths provide us with important clues to the history of genetics, since they reflect this understanding and record new developments as they were introduced.

> One constant rule in mythology is that *whatever happens among the gods above reflects events on earth*... Myth, then, is a dramatic shorthand record of such matters as invasions, migrations, dynastic changes, admission of foreign cults, and social reform. When bread was first introduced into Greece – where only beans, poppy seeds, acorn and asphodel-roots had hitherto been known – the myth of Demeter and Triptolemus sanctified its use; the same event in Wales produced a myth of The Old White One, a sow-goddess who went around the country with gifts of grain, bees, and her own young; for agriculture, pig breeding and bee-keeping were taught to the aborigines by the same wave of neolithic invaders. Other myths sanctified the invention of wine.[1]

The ubiquity and depth of interest in reproduction and inheritance in the past can be seen in the theories held by major cultures and civilizations, especially in their teachings about the domestication of plants and animals. After the emergence of agriculture and principles of breeding, people began to turn their new insights toward human reproduction. Elaborate myths grappled with such questions as how children are made and how the sex of a child is determined. We can look briefly at some of the early answers.

mythology and the domestication of plants and animals

Through their numerous drawings, carvings, and myths, early societies recorded the emergence of each useful crop and animal, as well as the societal impacts of these events. As they domesticated a crop or an animal that proved valuable to them, such societies often created a god representing or protecting it. By worshipping and propitiating the god with sacrifices, the people could control the whims of nature through that god while expressing their gratitude and their dependence on the plant or animal. The gods and goddesses created as protectors of each animal or plant symbolize their importance for the civilizations that produced them.

Nowhere are examples more abundant than in ancient Egypt. There, the domestication of cereals and grapevines proved so important to the well-being of the people that the great god Ousir (Greek name, Osiris) was given credit for their development. The Egyptians told stories of how he came to earth just to teach them to make ploughs, till the earth, plant seeds, and reap the harvest, and to introduce them to the delights of bread, wine, and beer. The handsome Osiris was a particularly pleasant god, gentle and full of songs, who wended his mythological way from Egypt around the world, spreading seeds and civilization as he went like a prehistoric Johnny Appleseed. His sister Isis, who was also his wife, was credited with teaching the women of Egypt to grind the corn that their husbands and sons grew, and to spin the flax and weave it into cloth. Women were probably history's first cultivators, who gradually domesticated edible plants and small animals, such as goats and sheep, while their husbands were preoccupied with the more primitive – and less productive, less reliable – ways of providing food and clothing. By the time of the Isis–Osiris myths, conventional domestic patterns had begun, and the intricacy of the myths indicate how completely Egyptian culture had developed the use of plants.

A number of animals, especially cattle, play important roles in the Isis–Osiris myths. It is hard to judge where cattle were first domesticated. The earliest cave pictures suggest that domestic cattle originally derived from three ancestral species: the aurochs (*Bos primogenius*) and longifrons (*Bos longifrons*) of Europe, and the zebu (*Bos indicus*) of India and Africa. Neolithic art from the Atlas mountains depicts domesticated cattle being led meekly about with ropes. The Egyptians had domesticated cattle before the Isis–Osiris myths were formulated, since Isis was associated with the cow – her sacred animal – and became identified with the cow-goddess Hathor.

Cattle domestication must have been long and difficult, since the original wild cattle were large and fierce. Individual animals were probably captured and penned up inside strong fences. Inbreeding of these artificially selected animals exaggerated certain characteristics with each generation until strains different from the original stock gradually emerged. Perhaps selection of some for religious sacrifices hastened the domestication; sacrificing the largest, fiercest, most imposing animals to the gods left the smaller, quieter individuals to reproduce, thereby gradually removing wild characteristics from the herds. People may have first domesticated cattle rather unwittingly and then later realized the advantages of having quiet cows that could be milked and herded easily, leading to further deliberate domestication.

The relatively docile, productive animal that resulted was so valuable to ancient societies that both male and female of the species were connected with the gods. The myths often dictated actions by the priests which stimulated further observations of breeding patterns. Wherever a domesticated animal (be it bull, cow, cat, or dog) became sacred to a god, the ancient Egyptians had to learn new lessons in genetics in order to select and breed suitably marked animals – to keep animal deities installed in the temples. For example, Ptah, a god of Memphis, was thought to have inseminated a virgin heifer and had himself been born again from her as a black bull, Hapi (or Apis, in Greek). Hapi, Ptah's incarnation, was kept in the temple of Ptah at Memphis. When the bull died, the priests had to replace him, and not just any bull would do. Hapi had to be black with a white triangle on his forehead and white markings on his back, right flank, and tongue, representing a vulture, crescent moon, and scarab, respectively. And the hairs of his tail had to be double.

To ensure that they had a suitable bull to replace a dying Hapi, the priests bred promising bulls and cows to maintain a herd of white-marked black animals. Each experiment carried out in the temple yards taught the priests more principles of selection and breeding, so knowledge of reproduction and inheritance began to increase rapidly. Thus, domestication led to veneration, which in turn led to active experimentation.

The mythologies of other civilizations also record developments in plant cultivation and animal breeding. The relationship between sex and reproduction was well understood in western Asia before the time of classical Greece. The Babylonians and Assyrians knew in 5000 BC that there are male and female date palms, and artificial pollination has been carried out since at least the time of King Hammurabi in 1790–1750 BC. In many reliefs from King Ashurbanipal's time (870 BC), priests wearing masks and wings to represent winged spirits use pine cones dipped into golden handbags of pollen from male trees to fertilize the flowers of the female plant. Obviously they recognized that the male and female forms were separate plants which require fertilization. This artificial pollination led to the development and proliferation of numerous new varieties of cultivated dates, now more than five thousand named varieties.

Meanwhile the ancient Chinese used their knowledge of genetics to breed surpassingly beautiful roses five thousand years ago. Roses became so popular that the Emperor Han had to destroy many of the gardens to allow room for more practical food production. The two main Chinese varieties, *Rosa chinensis* and *Rosa odorata*, were not introduced into Western stocks until the eighteenth and nineteenth centuries; but frescoes in Knossos show that roses (*Rosa gallica*) were cultivated in Crete by 1600 BC and two species (*Rosa gallica* and *Rosa damascena*) featured in Egyptian paintings and textiles a thousand years later. Mutant roses were selected and bred in antiquity as they are today; King Midas, for example, had gardens of sixty-petalled blooms.

Agricultural deities are the oldest of the Greek and Roman gods, and many were replaced by less peaceful deities as the civilizations became more warlike after domestication had become commonplace. The Greek god Pan survived; half man and half goat, the protector of shepherds and flocks, he was credited with making the ewes and goats fertile and prolific breeders, and earned a roguish reputation himself. He was revered for bringing agricultural civilization to Greece; he was reputed to have taught humankind the art of bee-keeping and the cultivation of the all-important olive trees and grapes.

An imaginative genetics, widespread in Greek mythology, reflects the state of understanding of reproduction the society had reached. The Greeks knew that each species *breeds true* – that is, produces only offspring like itself – and also recognized that each animal shows a combination of characteristics inherited from its mother and father. However, like people of many more modern societies, the early Greeks did not understand the barriers to interspecies breeding and often credited unfamiliar animals (such as the giraffe) as the offspring of parents of different species (such as the leopard and the ostrich). The genealogies of the gods reflect this belief. To explain the goat–man chimera Pan, they told how the Arcadian god Hermes approached Penelope in the form of a male goat and mated with her. Similarly, Pasiphaë, wife of Minos, was thought to have mated with a bull and so to have born the famed Minotaur, the half-man, half-bull terror of the Labyrinth at Knossos.

Like goats and cattle, the myths tell us, horses had been bred for domestic use in early Greek times. When Homer composed the *Iliad* about 900 BC, animal breeding was of considerable importance, as the description of a certain breed of horses owned by Aeneas of Troy shows:

> They are bred from the same stock that all-seeing Zeus gave Tros in return for his boy Ganymedes; and they were the best horses in the world. Later Prince Anchises stole the breed by putting mares to them without Laomedon's consent. The mares foaled in his stables, and of the six horses that he got from them, he kept four for himself and reared them at the manger, but he gave these two to Aeneas for use in battle. If we could capture them, we should cover ourselves with glory.[2]

And in Roman times, about 100 BC, Virgil understood the principles of breeding practices for horses and cattle and gave this advice:

> When the lusty youth of thy flock endures, let loose the males, put thy herds early to breeding, and generation after generation keep up the succession of thy flock.[3]

The great importance of horses is indicated by the remarkable ancient carvings in the chalk hillsides of England.

The myths of these diverse ancient peoples show clearly how vital a role the gradual understanding of plant and animal reproduction played in the rise of civilization and the origin of the great myth cycles. The myths emerge as poetic, imaginative renderings of ancient knowledge, creatively combining primitive science and historic legends to make all knowledge coherent, and also providing psychological stability between the known and the unknown for the primitive peoples. As these civilizations became more sophisticated, thinkers such as Aristotle began to differentiate between imagination and fact, and science began gradually to separate off from superstition. The first attempts to explain human heredity were far from scientific, but they reveal a deep conviction that its laws are ultimately comprehensible, and a passionate and lasting curiosity in the attempts to decipher these laws.

heredity in human society

The striking resemblance between children and their parents is a commonplace observation. Ancient people undoubtedly recognized that people could resemble each other if they shared a remote ancestor, and placed great emphasis on kinship. Underlying the concept of kinship, which provided cohesion in developing societies, the recognition that "blood is thicker than water" carried an implicit assumption about heredity within families. Ancient stories emphasize again and again that ancestry is a vital factor in determining one's characteristics; the long catalogues of "begats" in the Old Testament identify families and support men's claims to honor by relating them to revered ancestors. Environment was considered of secondary importance; strangers who could show blood ties were considered to have more in common than people who lived side by side. Few rituals were more honored by North American peoples than the ceremony in which unrelated friends symbolically proclaimed themselves "blood brothers" by mingling their blood in small self-inflicted wounds.

Kinship has always given society both a spatial organization and a temporal stability by organizing individuals into groups of relations, or families:

> The basic unit of ancient Hebrew society was the household (beth)… A number of related households constituted a clan (mishpachah); a

> number of related clans constituted a tribe (shebet); the twelve tribes constituted the nation (am). All of these were regarded as extensions of the family and the whole people was united by a sense of kinship.[4]

People derive a sense of order from knowing their place in such a structure. In recognizing the inheritance of physical, mental, and behavioral characteristics, people found a link between past, present, and future. Such a link gave vital continuity to existence, a sense of identity beyond the ephemeral present, and it provided our only link to earthly immortality.

Moreover, knowledge of heredity justified early structures, including social classes, that were necessary for societal stability. People assumed that capabilities and traits of character, as well as physical characteristics, were inherited, so it made sense for children of rulers to inherit their fathers' positions and for members of other classes to remain in their place. Children of the wise appeared to inherit their parents' intelligence; the offspring of craftsmen expressed talents for their fathers' trades. Servants were thought to bear children fitted for the same work as their parents (an idea still debated today). Priesthood was hereditary among the Jews of the tribe of Levi, like shamanism in the clans of the Siberian peoples; soothsaying ran in families among the ancient Greeks, and wealth and power were almost everywhere passed on from heir to heir. In India, the concept of caste turned the primacy of heredity into a rigid political system; in the Middle East and North America, slavery was a hereditary condition; and around the world the divine right of kings, chiefs, emperors, and tsars to hold dominion over their subjects was passed on to each generation through the act of intercourse. People felt that society was following the model of nature when its roles were filled without interruption by successive generations.

The birth of a child with a disease for which there is no hope of recovery has always represented an enormous physical, emotional, and economic drain on a family. The Vedas (*c.* 2500 BC) and Sutras (500–200 BC) of the ancient Hindus, for example, reveal an awareness that some illnesses can be inherited. They advise a young man about to choose a wife to check her family history, sometimes for up to ten generations, in order, according to the Astangasamgraha, "to make sure that she has no illness which could be inherited and that her family is free of such illnesses." This advice shows the knowledge that an inherited defect may skip a generation, so a healthy woman might still carry a hidden trait that could manifest itself in her children or grandchildren. These writings were not agreed on *which* diseases were inherited, which is not surprising, since communicable diseases and environmentally induced problems could both run in families. Nevertheless, the Manu Code of Law suggests rejecting families with histories of leprosy and epilepsy (as well as dyspepsia, consumption, and even hemorrhoids). But tendencies to good character and admirable deeds were also thought inheritable: the writings exhort the youth choosing a mate to ally himself to a family of good reputation.

Even curses were thought to be hereditary. The story of the doomed House of Atreus of Greek legend, preserved in the plays of Aeschylus and Sophocles, records its destruction due to Atreus's murder of Thyestes' children. The Hebrew God continued to punish transgressions into the third and fourth generations. Christianity adopted this conviction of the hereditary nature of guilt in the concept of original sin, according to which Adam's descendants are all guilty of his disobedience to God.

At least one genetic disease, Huntington's chorea, has resulted in the belief that witchcraft is hereditary. The fatal disease entails gradual physical deterioration, increasingly spasmodic jerking and weaving movements, and mental aberrations such as loss of memory, character change, and sometimes a tendency to violence. It appears only in middle age, usually after a person has married, had children, and passed the defect on. Perhaps the best-known victim of this disease is the folk-singer and songwriter Woody Guthrie, who died of Huntington's chorea in 1967. His wife and his son Arlo Guthrie – who is apparently free of the disease – have done much towards funding research and bringing public attention to the illness and the search for its cure.

In the United States it can be traced back for 350 years to a single family that emigrated from the English town of Bures in Suffolk to the Watertown Plantation in Massachusetts. The people of Puritan Massachusetts (as in England) were constantly on guard against witchcraft and intent on following the Bible's edict that witches must be put to death. The famous Reverend Cotton Mather claimed that witches blasphemously aped Christ's agonies on the cross. Since women with the spasmodic jerking of chorea fitted the description, Mather claimed that their ancestors had been mockers at Christ's crucifixion, so God had cursed their lineage with disease. One of the emigrant women returned to England, where she was hanged for witchcraft. Another, Elinor Knap, was exhibited at the prison-house in Connecticut for the crowds to witness her "witchlike" behavior before her execution in 1653. In 1671, Mather called for her daughter, "the Witch of Groton," to be hanged. In 1692, a Mary Staples, who was apparently Knap's sister, and Staples's daughter and granddaughter (Mary and Hannah Harvey), were tried for witchcraft under the Connecticut witchcraft law, which decreed that a child of a suspected witch could be held because "a witch dying leaveth some of ye aforesaid heirs of her witchcraft." Although the women were acquitted amidst charges of mischief against their accusers, even today Reverend Mather's legacy remains.

how are children made?

The profound role of inheritance in human cultural psychology, and the universal desire for healthy children that it fostered, led to a fascination with its processes and a desire to control it. From their knowledge of reproduction in plants, the ancients reasoned that reproduction begins when a male plants a kind of "seed" in a woman. Early Egyptians, for example, understood reproduction only partially and still saw the process more as a miracle than as a biologically comprehensible phenomenon. They credited this miracle of conception to Ra, the sun god, and believed that the god himself (in the guise of the Pharaoh) inseminated the Pharaoh's wife to produce the succeeding kings, who were therefore believed to be divine. The Egyptian king Akhenaton (Ikhnaton, 1379–1362 BC) celebrated the god's powers in his "Hymn to the Sun":

> Thou art he who createst the man-child in women,
> Who makest the seed in man
> Who giveth life to the son in the body of his mother,
> Who soothest him that he may not weep,
> A nurse even in the womb
> Who giveth breath to animate everyone that he maketh.

The Greeks appear to have been the first to make a concerted effort to understand the physical processes of reproduction. By the time Homer wrote the *Iliad*, the Greeks took for granted that physical features and traits of character are inherited, and the origin and lineage of the hero were considered vitally important. Homer recorded pedigrees only through the *male* ancestors, the forefathers; "foremothers" are relatively ignored. This cultural emphasis of a patrilinear society reflects the poet's preoccupation with the heroic virtues of courage and strength. But the early Greeks seem to have actually believed that hereditary characteristics come only from the father; rarely is any mention made of physical or behavioral resemblances to the mother. "The noble spirit of the father/Shines forth in the nature of his son," wrote Pindar in 446 BC, and his contemporary Euripides echoes the belief: "A noble father sires a noble son;/A base man's son is of his father's kind." Perhaps this belief was based on the observation that males produce semen during intercourse and females do not.

However, Alcmaeon of Crotona, a sixth-century BC physician and the first person to theorize on the physiological nature of semen, recognized that children often resembled their mothers, so females must also contribute to heredity. He surmised that women must also produce semen that remains internal and therefore invisible. (The sages of ancient India held a similar theory.) Alcmaeon decided that such a significant fluid must originate in the brain and flow from there to the genitals. Hippo of Rhegium, however, asserted that semen was formed in the spinal cord.

Later Greek philosophers, including Leucippus, Anaxagoras, and Democritus, noted that virtually all parts of the body exhibited hereditary differences. Therefore, they concluded, semen must be drawn from each organ and part of the body, and carried via the blood at the moment of copulation to the genitals. Plato (429–347 BC) agreed, and this theory – which later became known as *pangenesis* – remained current for centuries. (In the nineteenth century, Charles Darwin thought about heredity through a model of pangenesis, and this kept him from developing a genetic basis for his theory of evolution through natural selection.) Hippocrates (460–357 BC) accepted pangenesis but believed that hereditary traits were borne in the four bodily fluids, or humors – the blood, phlegm, black bile, and choler (yellow bile) – and that semen is drawn from these humors as well as from the organs. Anaxagoras remained convinced that females produce no semen and believed that a *homunculus* – a tiny, fully formed human – must already exist in the male semen when it enters the mother. This theory, too, persisted well into the nineteenth century.

The great philosopher Aristotle (384–322 BC), with his unique capacity to find the flaws in any argument, clarified early thinking about heredity along with many other subjects. Aristotle agreed that females do not produce semen but pointed out that, since they do contribute inherited traits, the homunculus theory is inadequate. He also criticized pangenesis with several astute observations: that it does not account for a soldier who loses an arm in battle nevertheless producing a normal child; nor for characteristics that develop in the parent after procreation (such as premature graying); nor for the transmission of non-concrete characteristics such as voice, posture, or way of walking; nor for traits that skip a generation, so that a child resembles a grandparent more that its parents. He noted that offspring often resemble their ancestors who, after all, contributed nothing to the semen.

Having devastated the pangenesis concept, Aristotle proposed that semen is formed from the blood. Since females produce no semen, he believed that their menstrual blood bears hereditary material. But "material" is the wrong word; Aristotle concluded that semen transmitted, not generative parts of organs, but a kind of non-material information – *a capacity for form* – which gave the developing embryo the *potential* for inherited characteristics rather than the characteristics themselves. This concept, formulated over two thousand years ago, is remarkably similar to modern genetic theory.

Aristotle wondered how this information could be organized during fetal development and how each organ could be placed in its correct position. He theorized that the first organ to form in the womb must be the heart, which already has a soul – a power that controls the organization of the total fetus. He suggested that children resemble their parents if all is normal in the womb, but that an abnormal relationship between semen and menstrual blood can diminish the resemblance.

Aristotle shared the common Greek belief that characteristics acquired during a person's lifetime could be passed on, so a scar or mutilation could appear in one's children. Although he knew that such inheritance does not always occur, he thought his concept of the transmission of potential, rather than of the actual characteristic, explained this discrepancy. As an example, he cited "a man who had been branded on the arm had a child who showed the same branded letter, though it was not so distinctly marked and had become blurred." The theory of acquired characteristics proved lastingly popular and has been propounded in increasingly ingenious discussions right up to the present.

Aristotle's ideas dominated thinking for centuries after ancient Greek and Roman times. Ovaries were discovered in 300 BC and correctly diagnosed as the female equivalent of testes. But even though Greek physicians and philosophers had gone a long way toward explaining inheritance in rational, biological terms, the public at large was ignorant of their theories and continued to devise its own imaginative explanations. They ascribed differences and resemblances between children and their parents to such factors as the mental images of parents during coition, thoughts suddenly passing through their minds, and abrupt changes in their moods.

After the fall of Rome, the knowledge of the Greeks and Romans was lost to all but a handful of scholars, and persisted through the Middle Ages only among Arabs such as the great philosopher-physician Ibn Sina (Avicenna, 980–1037) and the philosopher Ibn Rushd (Averroes, died 1198). As the Arabs retreated from Spain, their writings were translated into Latin by the Spaniards, and the Greeks' knowledge about reproduction was immediately recognized as a major challenge to Christian philosophy. (It was not until 1251 that Pope Gregory IX allowed the scientific works of Aristotle to be circulated.) Attempts to synthesize scientific and religious thinking became a high priority among theological scholars, but Aristotelian empiricism and Christian faith seemed frustratingly incompatible, and by the thirteenth century the now familiar gulf between science and religion had become firmly established. Among the great naturalists of medieval times, Albertus Magnus (*c.* 1200–80) accepted Hippocratic pangenesis but did not believe in female semen. His contemporary Thomas Aquinas believed that children resemble only their fathers and that female children were aberrations. These naturalists remained Christian but refused to accept those dogmas of the church that experimentation disproved. Roger Bacon (*c.* 1214–49), who agreed with pangenesis but believed semen to be formed from excess nutrients, was adamant in separating science from church teachings and met with powerful theological persecution. Later, Leonardo da Vinci (1452–1519) recorded Aristotle's theories without criticism and asserted that a mother and father make equal hereditary contributions to a child's characteristics. The Swiss alchemist Paracelsus (1493–1541) tried to reunite science with religion and philosophy, and devised his own version of pangenesis, relying heavily on the Hippocratic theories.

By the sixteenth century, educated laymen such as the French essayist Michel de Montaigne (1533–92) had become aware of classical theories of genetics and were intrigued by the questions they raised. Trying to fathom how he could have inherited the propensity to develop kidney stones from his father, Montaigne wrote an essay "On the Resemblance of Children to Fathers," in which he noted that his father did not suffer from this affliction until his sixty-seventh year, and that he himself was born twenty-five years earlier, when his father enjoyed his best health. Where, he wondered, was the propensity of this defect hatching all the time? When his father was so healthy, how did the sperm with which he made his son carry so great an impression? And why was Montaigne the only one of his many brothers and sisters to suffer from kidney stones? Montaigne challenged pangenesis, deducing, like Aristotle, that he must have inherited not the kidney stone but the *tendency* to produce these annoying objects.

In the seventeenth century, the Englishman William Harvey deduced that women must form eggs in the uterus and that these eggs must be fertilized by the male semen for a child to be produced. Soon the Englishman Robert Hooke and the Dutchman Anton van Leeuwenhoek put microscopy to good use. Leeuwenhoek examined human semen and the semen of other animals under his microscope and discovered spermatozoa. He believed, like Anaxagoras over two thousand years before him, that a tiny homunculus hides in a spermatozoon, a miniature child who gradually grows in the womb until it is ready to be born. Pierre Dionis, however, thought numerous sperm must be necessary to fertilize Harvey's hypothetical egg, since nature could hardly be so inefficient as to waste the millions of sperm present in each drop of semen.

During the eighteenth century, with its conviction of the orderliness of nature and the essential comprehensibility of its guiding laws, scholars by the dozen turned their attention to the intriguing question of how a child is made. *Spermatists* followed van

Leeuwenhoek and believed that each sperm cell contained a complete individual (homunculus). The Italian Marcello Malpighi disagreed; he began the alternative *ovist* school by proposing that the female egg (whose very existence remained quite hypothetical) contained a tiny homunculus, which was simply "awakened" when the egg was penetrated by a sperm cell. The battle between the two schools of thought raged for decades.

The Frenchman de Maupertuis suggested that the seminal fluid of each parent contains "particles" (*élémens*) which form a hereditary pool in which some of the particles link up to form the different parts of each fetus. Such a fetus therefore resembles both its parents. Extra particles cause monstrosities (as we shall learn, children with Down's syndrome, or mongolism, are now known to carry extra bits of hereditary material), as do deficiencies of particles. He stated that the leftover *élémens* remain in the system and can express themselves in later generations, thereby causing a child to resemble his grandparents or great-grandparents.

This theory begins to look more towards the tenets of twentieth-century genetics than back to the ancient theories of Hippocrates and Aristotle. Finally, in 1827, Karl Ernst von Baer discovered the mammalian egg, and the way was paved for the beginning of modern genetics.

chapter three

what is inherited?

What does it mean to say that an organism inherits its features? Here, for instance, is a family in which the mother has strikingly red hair and green eyes, while the father has brown hair and brown eyes; one of their children has very red hair, while the other has auburn hair, and their eyes are shades of brown or hazel. Both parents have earlobes that are not closely attached to their cheeks; one of their children, however, has lobes that curve downward and attach to his cheek. Both parents are taller than average, and their children are both growing taller than others of their age. The children have obviously inherited their features from their parents, even if an odd feature, such as the shape of an earlobe, shows up occasionally. But even before we can ask how this inheritance can occur, we must ask what exactly these features are.

Color in all plants and plants, including humans, is generally produced by chemical compounds called *pigments*. A pigment absorbs specific parts of the rainbow spectrum of light and lets other parts through; we see these transmitted parts of the spectrum as specific colors, whereas the full spectrum appears to us as "white" light. Black, brown, and red colors in humans are all the result of pigments. However, some colors, such as the blue color of human eyes and the iridescent sheen of the feathers of some birds, are not created by pigments but rather by the light-bending and light-scattering properties of other constituents.

What about height? Many factors determine how tall people grow, but among them are other chemical compounds such as hormones, especially one called human growth hormone (HGH). What about earlobes? In this case we have little idea why a fold of skin in the developing outer ear grows into one shape rather than another, but we can begin to understand it by seeing that the skin is a tissue made of many cells, so its shape depends on the way these cells grow and how they cling to one another. So the common feature of all these human features is that they involve specific chemical structures: pigments, hormones, and cells, which are complexes of many kinds of chemicals. Children inherit the features of their parents because they receive some kind of *instructions* from their parents, instructions that tell their bodies to produce specific pigments; to produce specific levels of HGH; to make their skin and muscle tissues grow in certain shapes resembling those of their parents, rather than those of other people.

> Inheritance means passing on instructions to make specific structures.

Although we are a long way from understanding how all these complex features are made, the direction of modern genetics is to seek this detailed understanding: to learn how the factors of heredity, which we call genes, can cause certain pigments or hormones or tissues to develop. And we can only understand this by first learning just what those structures are.

cellular structure

Just as the telescope revolutionized astronomy, the microscope revealed some of the most profound truths about the organization of life. We can only imagine the surprise and excitement scientists such as Anton von Leewenhoek felt when they glimpsed the previously invisible world of living creatures in a drop of pond water or a bit of soil. Leewenhoek described "animalicules" in semen and blood which were obviously alive. In 1665, Robert Hooke looked at cork shaved thin enough to transmit light and observed regular rows of boxlike structures, which he called *cells*. The structures Hooke saw were really just the walls of long-dead cells, but other observers began to realize that plants and animals are composed of similar units, of many shapes and sizes, filled with various structures that must be carrying out the functions of life. In 1839, Matthias Jakob Schleiden, a botanist, and Theodor Schwann, a zoologist, advanced the hypothesis that all animals and plants are aggregates of cells. They proposed that all living organisms begin as single cells and that multicellular creatures are formed by repeated cellular divisions. One of the most important principles of biology today is that all organisms are either single cells or complexes of cells, and that a cell is a basic biological unit, a structure bounded by a membrane that separates the interior from the environment. The cell is the minimal unit of life, and living things may be defined as things that are made of cells.

Figure 3.1 shows two cross-sectional cuts, one made through a plant stem and the other through a small worm. It is easy to see that these structures are both made of many small boxlike cells packed closely together. The cells in each organism are not all alike; they form several different kinds of *tissues*, such as epidermis (skin or outer layer) and muscle in animals, or wood in plants; each tissue is made of distinctive types of cells. But we recognize the cellular structure in general by at least two features: a definite *boundary*, which often gives a cell a boxlike shape, and a *nucleus*, a large round structure usually lying near the center of the cell (or near one side in many plant cells). A cell's complex contents are separated from their surroundings by a very thin *membrane* that keeps the cell intact and creates entrances and exits, which determine what can get in and out.

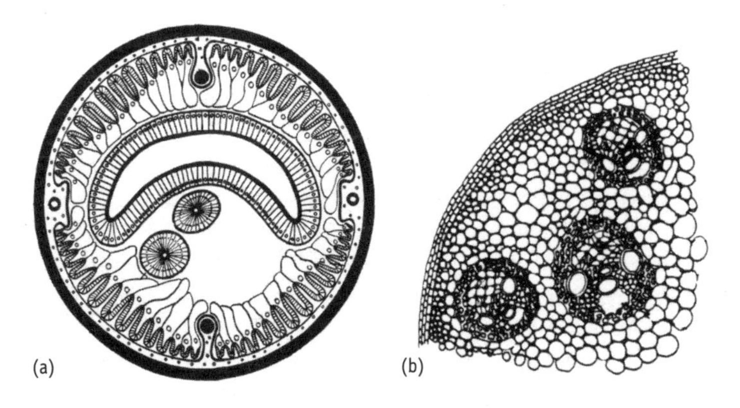

Fig. 3.1. *Thin cross-sectional slices viewed with a microscope show that complex organisms are made of many boxlike cells. (a) A small worm, (b) a plant stem.*

A single cell also exhibits the most fundamental form of reproduction. It requires the proper conditions, which may occur in the rich fluids inside our bodies, in the juices of a plant root, or in an artificial mixture of nutrients in a laboratory flask. Then each cell takes in nutrients from its surroundings and transforms them into more of its own structures, so it grows larger. Then eventually it divides in two:

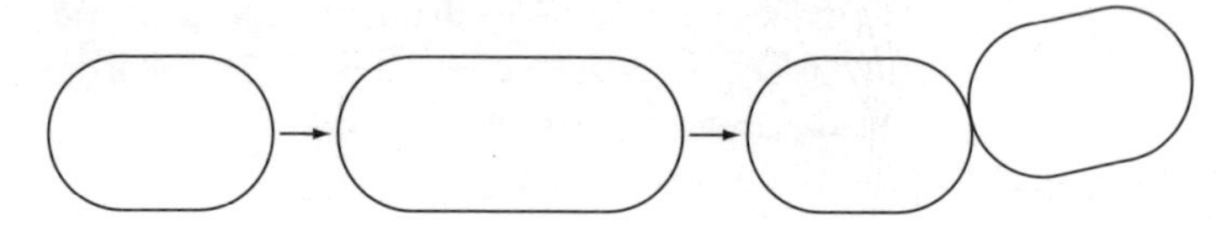

All organisms are divided into two groups based on cell structure. *Procaryotes*, mostly what we call bacteria, are extremely small and do not have a nucleus (see below). *Eucaryotes*, which include plants, animals, and many single-celled organisms such as amoebas and algae, do have a nucleus. We will focus for a while on eucaryotic cell structure.

Modern microscopes, especially electron microscopes, have revealed that cells commonly share many internal structures, called *organelles*, as shown in Figure 3.2. Their most prominent organelle is often the nucleus, a central structure bounded by membranes forming a nuclear envelope; the nucleus is especially important for genetics because it contains the *chromosomes*, which carry the stuff of heredity. The cell also has many small, elongated bodies known as *mitochondria*, the primary structures that derive energy from food molecules, such as sugars, and store it in a chemical form that the cell can use for all its activities. Many plant cells have brilliant green *chloroplasts* that capture the energy of light and store it in the same chemical forms. Both kinds of organelles are made primarily of *membranes*, which are thin sheets of material. We commonly see a variety of other internal membranes forming rounded *vesicles*, which can store various materials that will be needed for special purposes. Many cells have extensive membrane systems called *endoplasmic reticulum* where proteins and other materials are synthesized and directed to their proper locations in the cell; some substances are prepared for export to the outside of the cell.

In addition to *multicellular* organisms such as plants and animals – those made of many cells – the biological world includes many simple

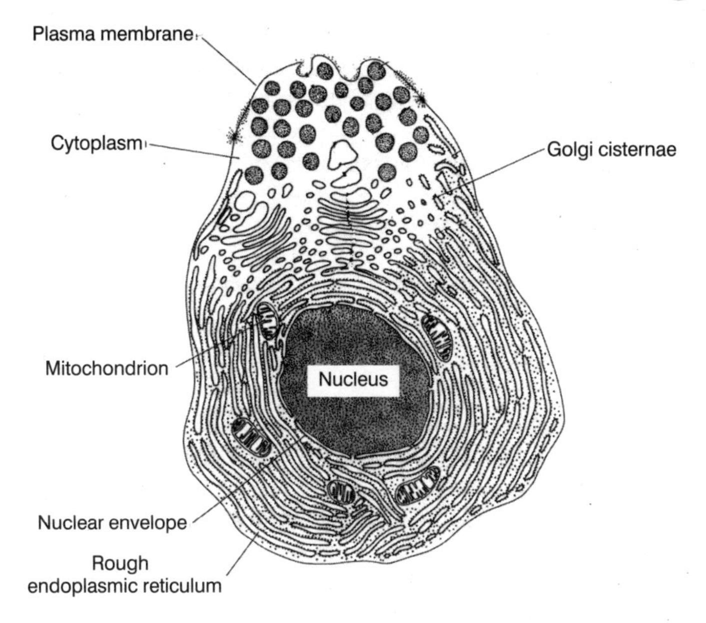

Fig. 3.2. *The cells of most organisms, including plants and animals, contain many substructures called organelles. This drawing of an animal cell shows what an electron microscope reveals, including the nucleus, several mitochondria, a series of membranes called the endoplasmic reticulum, and a layered bunch of membranes called the Golgi apparatus. The cell also contains many smaller structures, not visible here, that have specific functions.*

unicellular or colonial organisms. When viewed by microscopy, it is apparent that each one is a single cell, or perhaps a filament or cluster of many very similar cells. Some of these are algae and may have unusually shaped chloroplasts with brilliant colors. Others, called protozoa, whip around in pond or stream water by means of tiny hairlike extensions called cilia or flagella, or they may be rather formless amoebas that move by flowing and pushing out long fingers called pseudopods. The smallest of organisms, the bacteria, are tiny compared with the others: they may be only a tenth to a hundredth the width or length of larger cells (so their volumes are between a thousandth and a millionth the volumes of eucaryotic cells). Yet they are truly cells, for they have definite boundaries and regular shapes. They lack nuclei, however, because even though they have one or more chromosomes that carry their hereditary factors, the chromosomes are not enclosed in a surrounding nuclear envelope.

These common structures of all organisms, which we can see easily with microscopes, all have a more fundamental chemical structure, which we must understand to understand the matter of heredity. We will assume only the most basic knowledge of chemistry here: that matter is made of atoms, which unite in certain combinations into molecules, and that the formula of a compound tells the composition of its molecules. Thus, water has the formula H_2O because each of its molecules is made of two hydrogen (H) atoms linked to one oxygen (O) atom. You should also know that the atoms of every element have a definite weight (or mass): hydrogen weighs one unit,[1] carbon twelve units, and iron 55.85. The molecular weight of a molecule is the sum of the weights of its constituent atoms.

molecular structure

Here are two crystalline structures – diamond and calcite – that are quite typical of ordinary matter:

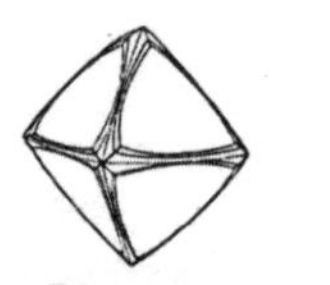

Diamond

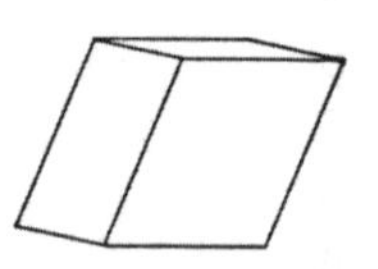

Calcite

Our world is largely made of materials like this. There must be a reason for their obvious regularity, and we begin to understand it by cutting them up into smaller pieces. Setting the diamond aside – since our research budget won't allow us to experiment with it – we can begin by carefully striking the calcite with a chisel and hammer. It breaks up into smaller chunks, and the remarkable thing is that each chunk has the same general form as the larger piece – not the same dimensions, but the same angles between all the edges and faces. We can go on breaking each piece into smaller ones, even if we have to use a microscope and a very fine chisel, and the same pattern is repeated. It is apparent that even pieces too small to see must have the same structure.

Calcite is known by chemists as calcium carbonate, and its structure consists of a carbonate group (CO_3 – a carbon atom joined to three oxygen atoms) plus an atom of calcium. Physical observations show that calcite has many carbonate groups and calcium atoms interspersed in a regular arrangement, creating at an atomic level exactly the angles of the large calcite chunk that we can hold in our hands:

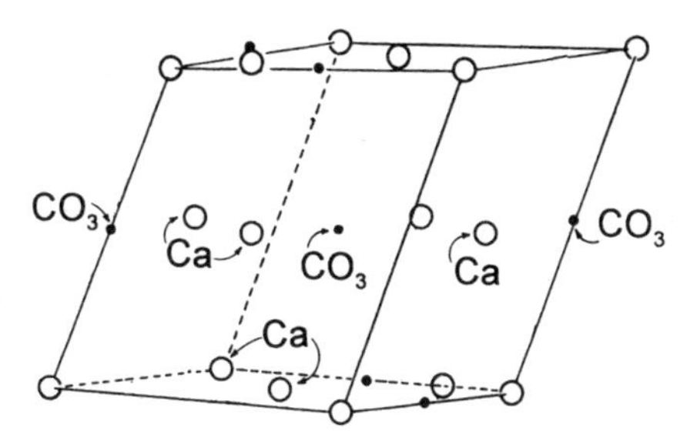

Thus, the visible structure of the material is the result of this crystalline structure extended and amplified many times over.

> Large-scale structure results from regularities in small-scale structure.

The structures of living materials are determined by their minute molecular structures. Many biological structures look quite crystalline, and the microscope reveals their beautiful, regular shapes. Tissues, we have seen, have remarkably orderly arrangements of cells. All this depends upon the materials they are made from.

The cells and tissues of all organisms are all made of the same general kinds of material. First, *water*. Water constitutes such a large part – 70–90 percent – of most biological structures that the properties of water largely determine biological processes. Dissolved in this water are salts of such elements as sodium, potassium, calcium, magnesium, and chlorine. The remainder consists of *organic compounds*, which are made of carbon (C) atoms combined with hydrogen, oxygen, nitrogen (N), and sometimes sulfur (S) or phosphorus (P).

The simplest organic molecules, of the kind found in natural gas and oil deposits, are methane, ethane, and propane.

```
     H            H  H            H  H  H
     |            |  |            |  |  |
   H—C—H        H—C——C—H        H—C——C——C—H
     |            |  |            |  |  |
     H            H  H            H  H  H

  Methane        Ethane          Propane
```

They are called hydrocarbons because they consist of hydrogen and carbon atoms. We picture these atoms as minute balls that form *chemical bonds* with each other by sharing a pair of electrons – one contributed by each atom. Each bond between two atoms is represented by a line in our pictures. Every element has a characteristic *valence*, the number of bonds its atoms can form with other atoms. Carbon has a valence of four, so each carbon atom can be attached to four other atoms; this allows many combinations of atoms and an enormous variety of organic molecules (Figure 3.3). (A pair or a triplet of lines represents two or three bonds together – a double or a triple bond.) Bonds made by sharing electrons are *covalent bonds*, and they are very strong: it takes considerable energy to make them and to break them, so organic molecules are quite stable. On the other hand, when such materials are burned (oxidized), the bonds are broken, releasing a lot of their energy, so these molecules are valuable fuels.

In the simplest organic molecule, methane, the carbon is attached only to four hydrogen atoms. In the other molecules shown, the

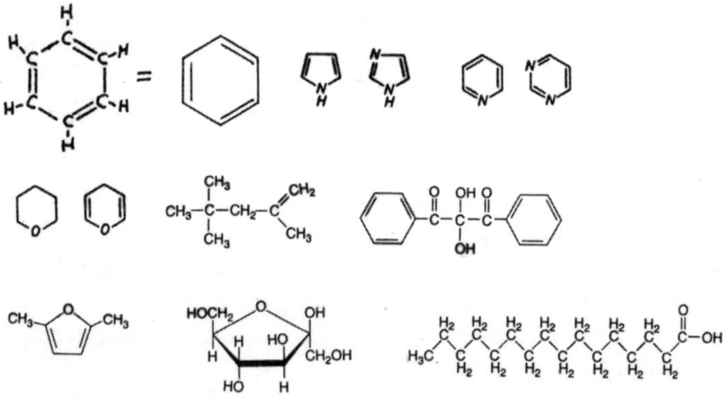

Fig. 3.3. *A variety of organic molecules, which are fundamentally built of carbon atoms, generally in chains. Each line between atoms is a bond, a pair of electrons shared by the atoms. Double and triple lines represent two bonds or three bonds between the atoms. The more complex molecules, especially those that have a ring shape, are usually drawn as simple lines, where each point (vertex) represents a carbon atom, generally with one or two hydrogen atoms attached (but not shown). Since the valence of carbon is four, each carbon atom must have four bonds; so if the drawing shows only three bonds on a C atom, there must be one H atom attached to it but not drawn in.*

carbon atoms share some of their bonds with other carbons, thus forming C–C chains with the other positions being filled in by hydrogen atoms. C–C chains can become very long; waxes may have chains of 30–36 carbon atoms. They can also be bent around to make rings of various sizes. But much of the variety comes from attaching groups of other kinds of atoms. For instance, OH (oxygen linked to hydrogen) is a *hydroxyl group*, and attaching it on a carbon chain creates an *alcohol*. An *amino group*, consisting of a nitrogen atom bonded to two hydrogen atoms (NH_2), can be attached to the chain to make an *amine*. In more complex groups, an oxygen atom is joined to carbon by a double bond (C=O), and one such combination, the *carboxyl group* represented by COOH, makes an acidic molecule. (An acid is any compound that can release a hydrogen ion, as the carboxyl group does, an *ion* being a positively or negatively charged atom.)

Combining these groups in all kinds of ways, on carbon chains of various lengths and on rings, makes an enormous variety of organic compounds. But only a few of them account for most of the structure of an organism. The most important compounds are proteins, nucleic acids, carbohydrates, and lipids.

The *lipids*, which are mostly what we know as fats and oils, consist of long hydrocarbon chains, usually with 16–18 carbon atoms. We are all familiar with their properties, for they are the sort of material that soils our clothes and can't be removed with water. Oil and water, says the old saw, don't mix. Materials that mix with water are said to be *hydrophilic* (literally, "water-loving"); those, like oil, that separate from water are *hydrophobic* ("water-fearing"). Fatty, oily dirt on clothing has to be removed by a dry cleaner, using solvents such as carbon tetrachloride or benzene–organic solvents that are also hydrophobic. Lipids, in fact, may be defined as materials that dissolve only in hydrophobic solvents.

The other important biological materials are remarkable because their molecules are huge. The small molecules – such as propane, benzene, or sugars like glucose – have molecular weights of a couple of hundred daltons at most. In contrast, proteins, nucleic acids, and some other structural materials of cells are called *macromolecules*, because they have molecular weights in the thousands, or larger. We should expect structural materials to be large. After all, humans make structures out of large, solid things like steel girders and sheets of plywood or sheetrock. The solid stuff of cells also consists of large components.

But all macromolecules have a fundamentally simple structure. They are all *polymers*, which are made by stringing together many similar or identical molecules called *monomers*:

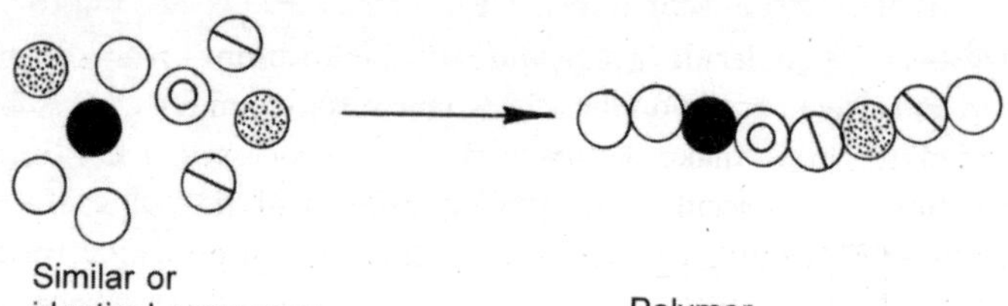

For example, the carbohydrates are all based on sugars, which are small organic molecules with a formula such as $C_6H_{12}O_6$. Those sugars of major interest to us – such as glucose, galactose, and mannose – have ring-shaped molecules. They can be linked to one another to make long chains, which may even contain branches. When glucose molecules are linked in one way (what chemists call beta 1:4 linkages), they form *cellulose*:

Cellulose is the tough fibrous material of plant cell walls, and therefore the principal component of wood. But when linked in a different way (alpha 1:4 linkages, with some 1:6 branches), they form *starch* and *glycogen*, which are food-storage materials of plants and animals. Other kinds of sugars, when linked in various ways, form the pectins and gums that constitute the fleshy substance of fruits and other plant parts. All these polymers, which may weigh many thousands of daltons, are called *polysaccharides*; their constituent monomers (sugars) are called *monosaccharides*. Other polymers are named similarly, using the prefix "*poly-*" (many).

An important class of polymers, the *proteins*, are long chains of monomers called *amino acids*, so named because each one has an amino (NH_2) group along with an organic acid (COOH) group. Two amino acids can be linked by joining the acid group of the first to the amino group of the next, with the removal of one molecule of water:

$$H_2N{-}CHR_1{-}COOH + H_2N{-}CHR_2{-}COOH \longrightarrow H_2N{-}CHR_1{-}CO{-}NH{-}CHR_2{-}COOH + H_2O$$

Peptide linkage

The resulting molecule (a *dipeptide*) still has an amino group on one end and an acid group on the other, and so a third amino acid can be added to it, to make a *tripeptide*. This process can continue indefinitely; a molecule made of many amino acids linked like this is a *polypeptide*, another name for a protein. Typical proteins have at least 200–300 amino acids joined in one long chain. (Lacking complete amino and acid groups, the part of an amino acid that remains in the chain is called an amino acid *residue*.) Since the average amino acid has a molecular weight of about a hundred daltons, a chain of three hundred amino acids has a molecular weight of about thirty thousand, a common size for a protein.

Real proteins are made of by linking twenty different kinds of amino acids, which differ only in the structure of their side chains (Table 3.1). They can be arranged in any sequence, so cells can form an enormous variety of proteins. The variety is way beyond human comprehension. Since there are twenty choices for the first amino acid in a chain and twenty for the second, there are 20×20 or four hundred ways just to make a dipeptide (with two residues). There are eight thousand kinds of tripeptides, 160,000 kinds of tetrapeptides, and thus 20^{300} ways to make a chain of three hundred amino acids. No one can imagine such a large number. The range of proteins made by all the organisms that have ever existed on earth constitutes only a small fraction of this number.

Every kind of protein has a unique sequence; for instance, part of the human hemoglobin A molecule, the red substance that carries oxygen in our blood, begins with the sequence Val–His–Leu–Thr–Pro–Glu–Glu–Lys–Ser–Ala–Val–Thr–Ala–, using the three-letter abbreviations for the amino acids. Every molecule of hemoglobin A in a normal person has precisely this structure. The simplest functioning organism has at least a couple of thousand different kinds of proteins, and a complex one, like a human, has on the order of thirty to fifty thousand. (Recent work on elucidating the human genome puts the number in this general range, but with a lot of

Table 3.1. Monomers of proteins, the amino acids

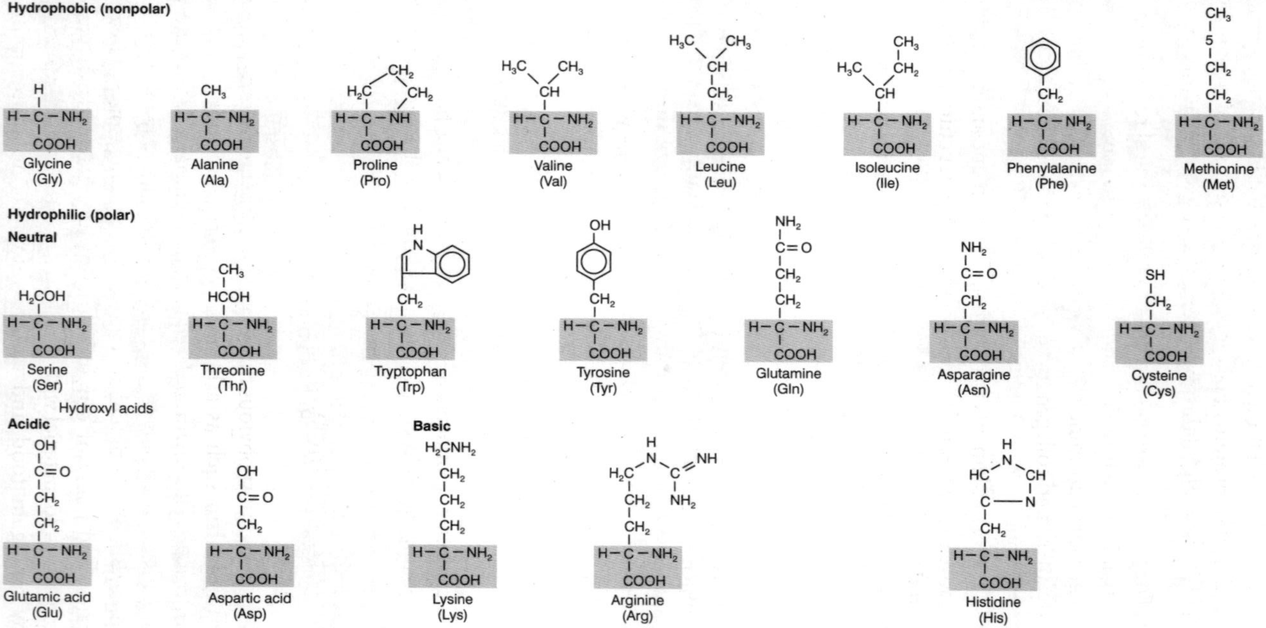

Hydrophobic (nonpolar)

Glycine (Gly) | Alanine (Ala) | Proline (Pro) | Valine (Val) | Leucine (Leu) | Isoleucine (Ile) | Phenylalanine (Phe) | Methionine (Met)

Hydrophilic (polar)

Neutral

Serine (Ser) | Threonine (Thr) | Tryptophan (Trp) | Tyrosine (Tyr) | Glutamine (Gln) | Asparagine (Asn) | Cysteine (Cys)

Hydroxyl acids

Acidic

Glutamic acid (Glu) | Aspartic acid (Asp)

Basic

Lysine (Lys) | Arginine (Arg) | Histidine (His)

uncertainty.) Each protein has a structure that is good for a different function, for proteins are the major workhorses of an organism. They do virtually all the important jobs we associate with life:

- They are the *enzymes* that make all the chemical reactions in an organism occur rapidly and in a controlled way.
- They form prominent structures: keratins make hair, skin, and feathers, and collagen forms much of the substance of cartilage and bone.
- They form filaments that push and pull on one another to create movement in muscles and other movable structures, such as cilia and flagella.
- They constitute an important class of *hormones,* which carry signals from one kind of cell in the body to another.
- They form *receptors*, which receive signals by binding to other molecules. A cell receives a signal from a hormone because the hormone molecule binds to one of its receptors. Receptors like those we use for tasting and smelling allow organisms to detect the presence of small molecules in their environment and respond to them.
- They are transporters that carry ions and small molecules across cell membranes, thus forming the basis of our nervous systems and organs such as kidneys.
- They are regulatory elements that control all kinds of processes so they occur at the proper rates.

We can begin to understand how cells are made and how they operate by looking more closely at some of the functions that proteins perform.

growth and biosynthesis

One of the most obvious things an organism does is grow. Growth of an organism such as a human being is the product of the two processes: cell growth and cell division. Humans, like many other organisms, grow only to a certain size and then stay more or less the same. But all our tissues are constantly changing, some at incredible rates, and many of our individual cells are constantly growing, replacing old material with new. It is obvious that an organism grows by taking in food and transforming it into more of its own characteristic structure. You are what you eat, says another old saw. We transform some of the molecules of our food into the substance of our bodies, plus wastes such as carbon dioxide, water, and urea. Photosynthetic organisms use such wastes as their "food" and transform them into the substance of their own cells. Either way, atoms from the environment become atoms in a growing organism.

Growth is fundamentally a chemical process. As molecules interact, their atoms exert forces on one another that break bonds and form new ones, so molecules can engage in *chemical reactions* in which their atoms are rearranged. For instance, with the aid of some heat, oxygen molecules in the air attack the carbon atoms in charcoal, and the two combine to make carbon dioxide (CO_2). This is what happens when charcoal burns, and heat is released because the new bonds, between carbon and oxygen atoms, contain less energy than the old bonds in the charcoal and the oxygen alone. Carbon dioxide molecules, in turn, can combine with water molecules to make carbonic acid: $H_2O + CO_2 \rightarrow H_2CO_3$.

Life depends upon such reactions, for each cell must take atoms and molecules from its surroundings and transform them into the materials of its own structure, while extracting energy from them. For instance, we must have a continuous supply of the sugar glucose in our blood because all our cells use that sugar for energy; they also maintain themselves and grow by rearranging its carbon, hydrogen and oxygen atoms into the complex structures of proteins and other macromolecules.

The process by which an organism manufactures itself takes place within cells and is called *biosynthesis*. Picture a cell as a manufacturing plant, but one that makes copies of itself rather than making cars or TV sets. A factory operates by means of *assembly lines*. A factory assembly line begins when someone puts together a basic part. The next person in line adds some small piece, someone else adds another piece, and so on until the whole item is complete. A complex product, such as an automobile, is made by several assembly lines, which make units that are finally assembled into the whole thing.

An organism synthesizes itself in the same way. All the chemical activity it conducts is known as *metabolism*. Each assembly line is called a *metabolic pathway*, and the molecules being transformed through metabolic pathways are *metabolites*. In each pathway, the metabolites are transformed step by step by adding or removing atoms or small groups of atoms, until they finally become the end-product. A metabolic pathway might be a series of reactions like this: first, two hydrogen atoms are removed from adjacent carbon atoms in a molecule; second, a molecule of water is added to those carbon atoms – H to one carbon atom, OH to the other; third, the hydrogen atom is removed from the OH, leaving the oxygen atom double-bonded to its carbon atom; fourth, an NH_2 group is added to another carbon atom. These changes must be made in very small steps because that is the nature of chemical reactions; every functioning cell contains hundreds of different metabolites, perhaps thousands.

These pathways are primarily directed toward synthesizing the monomers and other rather small molecules of a cell. (Some metabolic pathways actually *dis*assemble food molecules to release their energy, which is then used for biosynthesis and other processes; but we can ignore them for now.) Thus, the cell has pathways for making the twenty amino acids that constitute its proteins, each of the sugars that constitute its polysaccharides, each of its lipids, and so on. All the end-products of these pathways are then incorporated into larger structures such as proteins or cell membranes.

enzymes

An assembly line is operated by people, of course (although robots are replacing them increasingly in many factories), but what substitutes for them in a metabolic pathway? Just what effects the chemical reaction that transforms each metabolite into the next one? Sometimes nothing special is needed. Some chemical reactions occur spontaneously and rapidly when the appropriate molecules are mixed, just because the molecules bump into each other and have enough energy to react. Put some iron in contact with water and they spontaneously react to form rust, an iron oxide. But organisms can't depend on reactions occurring spontaneously. Many reactions simply can't happen unless some energy is added, and all cells have mechanisms for getting the energy to drive these reactions. Furthermore, almost all reactions of metabolism ordinarily occur too slowly, so there must be some way to speed them up. For this purpose organisms use *enzymes*. An enzyme is a protein that can interact specifically with some molecule – called its *substrate* – and cause it to undergo a certain chemical reaction; an enzyme can repeat the process with molecule after molecule, sometimes thousands of times per second.

Each metabolite has a specific shape. An enzyme that operates on it has a little pocket – its *active site* – that has a *complementary* shape (as pieces of a jigsaw puzzle do), so the small molecule can fit into it:

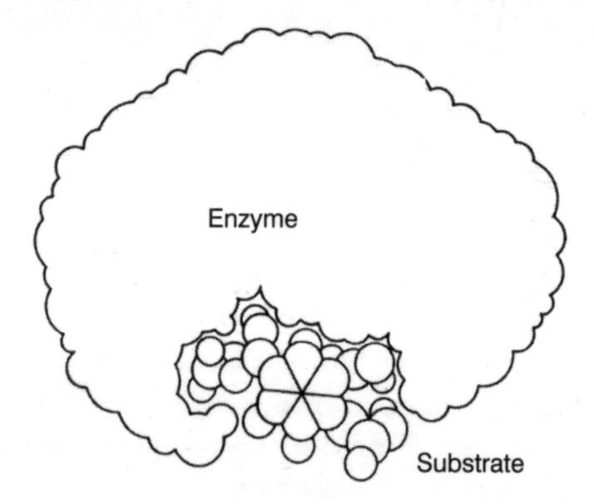

The features of the enzyme determine just what will happen to the substrate. Thus, enzyme A could attach a hydroxyl group to some metabolite, enzyme B could cleave the same metabolite in two, and enzyme C could remove an amino group from it. Each of these enzymes could function in a different metabolic pathway, for many metabolites are transformed into several different end-products.

Enzymes are the "workers" that operate the steps of a pathway. Some pathways are physically organized in a cell just like assembly lines, so molecules can move directly from one enzyme to the next. But many operate just because all the enzymes and substrates are mixed in a tiny compartment in the cell.

Enzymes, we have said, are all proteins. Figure 3.4 shows the structure of one enzyme, in which the chain of amino acids that makes its basic structure folds up in a specific way, forming an active site. In this site, certain amino acid residues hold their side chains in just the right position to interact with the atoms of the substrate and effect the right chemical reaction. The enzyme and substrate have been compared to a lock and key that fit together perfectly, but the two molecules really interact more dynamically, both changing their shapes as they combine and separate. An enzyme is such an excellent

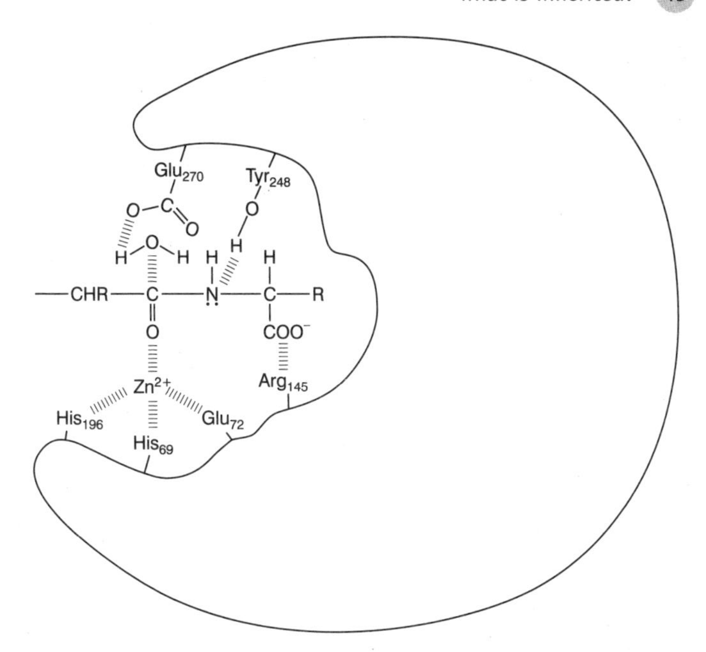

Fig. 3.4. *The structure of the enzyme carboxypeptidase, showing how amino acids (three-letter abbreviations with numbers) form an active site with the right shape to perform the one chemical reaction that this enzyme catalyzes. This is a digestive enzyme that cuts protein molecules in food. It uses a zinc ion (Zn). Dashed lines show molecules interacting.*

catalyst that it can drive a reaction thousands of times faster than it would occur spontaneously, and one enzyme molecule may operate on many thousands of substrate molecules per second. It is important to note that a cell has many molecules of each enzyme and large numbers of each kind of metabolite, which are constantly being transformed into one another. We say that the cell contains a *pool* of each metabolite (or end-product); some enzymes are constantly adding new molecules to the pool while others remove molecules from the pool and send them on their way along various pathways.

Other enzymes, or enzyme-like proteins, transport molecules across cell membranes. These membranes are thin layers of lipid and protein molecules (Figure 3.5(a)) that are virtually impenetrable by most small molecules and ions; transport proteins move them in and out of cells, or between two compartments of a cell. Figure 3.5(b)

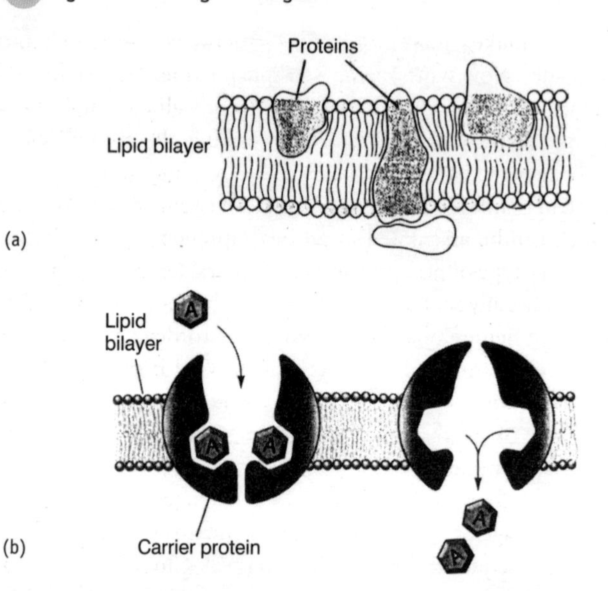

Fig. 3.5. *(a) All cell membranes are thin sheets of lipid molecules with many kinds of proteins embedded in them. All the molecules are free to diffuse rapidly through the lipids. Most molecules can't pass through the lipid layer unless they are fat-soluble. (b) Transport proteins have narrow channels running through them. They specifically bind to small molecules, or ions, on one side of the membrane and then change their shape to let these molecules move through to the other side.*

shows how one kind of transport protein may operate. A channel running through its center spans the membrane, but this channel has the same kind of specific shape as any active site, so it can bind only one kind of molecule. When this molecule binds on one side of the membrane, a contortion of the protein pushes it through the channel to the other side. Some transport proteins use energy to concentrate one kind of molecule inside the cell and another outside, thus controlling the composition of the cellular contents.

synthesizing polymers

Simple biosynthetic pathways produce all the amino acids, sugars, lipids, and other small molecules of the cell, which are mostly linked together to make macromolecules such as proteins and polysaccharides. Now synthesizing a simple polymer like cellulose is not complicated. Since cellulose consists of many glucose molecules, one kind of enzyme can synthesize it by linking glucose molecules into long chains, repeating the same process over and over.

But protein is different. A protein is made of twenty types of amino acids, which can be assembled in an incomprehensible number of ways. But every type of protein has its own characteristic sequence of amino acids. Just any sequence won't do. So a bone-marrow cell that is synthesizing hemoglobin needs some way to determine exactly which sequence of amino acids to assemble. What it needs, in other words, is *information*. Information is what you get when you specify one thing out of a range of possibilities. When you determine someone's phone number or find out how tall the Eiffel Tower is, you have acquired information because you know one correct number out of a wide range of possible numbers. Similarly, specifying that a protein has the sequence Ser–Gly–Ala–Ala–Val–Glu–His–Val– … is providing information. So every organism needs information to operate properly, and therefore it needs another kind of molecule: an information carrier. If humans consist of approximately fifty thousand types of protein, then our cells need specific instructions about how to line up the correct amino acids to make each of these proteins.

It should now be clear that this information must be the stuff of heredity, for it is just this detailed structure that makes each organism what it is. Each organism gets the instructions for its structures from its parents. What one generation must pass on to another are instructions for making each of its proteins, to make *this* particular pattern of protein (red hair, lots of growth hormone, earlobes not attached) rather than *that* one (blond hair, moderate growth hormone level, earlobes attached).

> Hereditary information is primarily needed to specify the structures of all the proteins of an organism.

To explain exactly where this information of heredity comes from would be getting way ahead of our story. That is the subject of the rest of this book. But these days almost everyone knows that the secret lies in DNA – deoxyribonucleic acid. Nucleic acids are another kind of polymer, quite different from proteins or polysaccharides, which we shall discuss in Chapter 7. The critical point is that DNA consists of four different kinds of monomers, which can be arranged in any sequence, so a DNA molecule also carries information in the sequence of its monomers, and a cell uses this information to direct the synthesis of its proteins. We will eventually define a *gene* as the DNA structure that carries the information for making a single type of protein. Furthermore, nucleic acids not only carry information but also can *replicate* themselves. That is, a DNA molecule can direct the synthesis of replicas of itself, so the information it carries can be passed on to each new cell or new organism. This is what heredity is all about.

cells as self-renewing, self-reproducing factories

Now think again about what an organism does. It takes raw materials from its environment and transforms them, through metabolic pathways, into the molecules of its own structure – first monomers, then polymers. But what is that structure? Primarily more of the enzymes that operate those very metabolic pathways, to make more of those monomers and polymers, which are mostly more of the same enzymes that operate those metabolic path... Well, you get the point. The organism consists of structures whose function is to make more of themselves. These are primarily proteins, and they receive information for their structure from nucleic acid molecules, generally deoxyribonucleic acid, or DNA. The total complement of DNA constitutes the *genome* of the cell. Looked at another way, the genome is the physical structure comprised of all its genes. Hereditary information, or *genetic information*, specifies how to make more of the structures of life, both the catalysts and the genome itself. Eventually, once a cell has accumulated enough new structures to roughly double in size, the cell divides, and then the whole process starts over again. Hence the fundamental unit of life (the cell) is merely a machine programmed with information for making copies of itself.

It is time to examine the details of how this is done.

chapter four

the breakthrough: mendel's laws

Heredity has always intrigued humans. In the first century BC the Roman philosopher Lucretius noted that children may resemble their grandparents or great-grandparents. A century later, Pliny the Elder wrote, "It often happens that healthy parents have deformed children, and deformed parents healthy children or children with the same deformity, as the case may be." From the beginning of agriculture, people recognized that certain traits of corn, sheep, or other domesticated plants and animals were inherited and hence could be modified by breeding selected types. Down through the ages, when people exclaimed, "He has his mother's smile" or "She has her father's temperament," they expressed an awareness of a hereditary basis for human characters.[1]

However, although hereditary transmission was known in early times, it was not understood. No attempt to find a simple pattern for inheritance was successful. We might try to understand the inheritance of human features with the obvious and simple idea that parental characters are blended in a child, so children would be an *average* of their parents' characters. Heredity would then be like mixing a can of red paint with one of white paint to produce pink paint, so a simple property like hair or eye color or the shape or size of a nose, and even complex behaviors, would reflect this blending. Then further mixing would never yield a paint color exactly the same as the original red or white. But even two thousand years ago, the Romans knew that isn't true. Another complicating factor is that a complex behavioral character such as mechanical ability is strongly affected by environmental conditions such as school and parental influences.

For Charles Darwin, the inheritance of characters was a key element in the theory of evolution. Just as farmers improve livestock by selective breeding, so natural selection favors reproduction of the fittest. If traits leading to a reproductive advantage were not inherited, evolution could not occur. However, Darwin seized upon the erroneous notion of pangenesis as the basis of heredity. As we have seen, this hypothesis suggests that miniature representatives of each part of our bodies (*pangenes*) collect in the gonads and are distributed into each gamete (sperm or egg). Thus, each gamete would carry a mixture of pangenes of toes, hair, teeth, and so on. This theory was the accepted basis for inheritance late in the nineteenth century, and it continues to color the thinking of people even today.

However, all the early ideas on the nature of the heredity were largely based on speculation. It was not until the middle of the nineteenth century that the experiments of Gregor Mendel provided data that he interpreted to show for the first time the correct basis for the transmission of characters.

mendel's discoveries

Gregor Mendel made the breakthrough that provided the key to heredity. In a monastery in Brno, now part of the Czech Republic, he studied the inheritance of certain characters in sweet peas. Yet the report on his results that he published in 1865 was met by indifference. Even though he had uncovered one of the keys to heredity and to the theory of evolution, which had excited so much attention only six years earlier, the few scientists who knew of Mendel's work didn't recognize its significance. It was only in 1900 that three biologists, using different organisms, obtained results verifying Mendel's, and he received posthumous recognition for his pioneering discoveries.

Why did Mendel succeed where so many others had failed? First, he studied only simple, clearly defined characters, such as flower color or seed shape. Not all characters are so clearly identifiable as being hereditary. Such characters as the height of a plant – or intelligence or the shape of the nose in humans – are so complex and variable that it is difficult to trace their transmission. Complex characters so well marked that they can be recognized as hereditary are rare. Mendel also followed the transmission of each character for several generations. Perhaps most important, though, he recorded the *numbers* as well as the appearance of the offspring, and he thought about these numbers statistically.

Classical genetic experimentation always uses two or more varieties, or *strains*, of the same kind of organism, which differ in simple characters such as flower color in plants or fur color in mammals. Mendel began with *purebred* strains of peas, strains that had been bred only with themselves for several generations so they consistently showed only one form of each character. Such strains are said to *breed true*. Each of his experiments was a *genetic cross*, or hybridization, of two such strains. A cross is performed by mating individuals of the two strains with each other to make *hybrid* plants, which might have some combination of traits that can then be followed in future crosses. In one case (Figure 4.1, p. 53), Mendel crossed purebred yellow-seeded with purebred green-seeded plants by removing the pollen-bearing parts of each plant and dusting it with pollen from the opposite type. In this simple representation, the × is read "by" and the arrow points to the progeny of the cross:

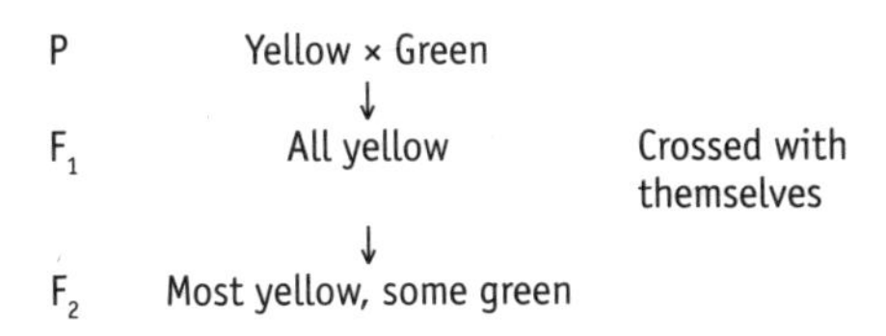

Though the peas resulting from this cross might have been yellow-green intermediates or a mixture of green and yellow, they all turned out to be yellow. It was as if the green color had simply disappeared in this generation, called F_1 for first filial generation (*filius* = brother). Mendel then planted the F_1 seeds and crossed the resulting plants with one another to make a second filial generation (F_2). Remarkably, the green color that had disappeared in the F_1 plants appeared again; some of the F_2 seeds were yellow and some were green. The other features Mendel studied showed the same pattern. So when he crossed a purebred purple-flowered strain with a purebred white-flowered strain, the F_1 had all purple flowers, but the white color reappeared in the F_2.

In contrast to earlier investigators, Mendel made a point of counting the plants (or seeds) with each trait. In following the inheritance of seed color, he obtained 6022 yellow and 2001 green seeds in the F_2. In following flower colors, he obtained 705 purple and 224 white plants in the F_2. These numbers look meaningless by themselves, and in similar situations, Mendel's predecessors threw up their hands and concluded that nothing could be concluded. Mendel, however, saw that both sets of numbers have a ratio very close to 3:1, and this observation led him to a simple explanation.

Mendel developed a *model*, a hypothetical picture of how his system might operate. The value of a model depends on how well it explains past observations and predicts the results of experiments that haven't been done yet. Mendel's model postulated that the plants carry "factors" that determine the inheritance of each character, and that every plant carries a pair of hereditary factors for each character, one factor derived from each of its parents. Furthermore, he postulated that when a plant has two different factors, one of them is *dominant* – its effect is made visible – while the other is *recessive* – its effect is hidden. Yellow seed color must be dominant, whereas green is recessive; purple flower color is dominant to white color. This feature of heredity is the basis for one common system for symbolizing hereditary factors: a capital letter for a dominant factor and the lower-case letter for the corresponding recessive factor. For instance, we use *Y* for the factor that specifies yellow seed color and *y* for one that determines green. Today, we realize that these two factors are forms of a single gene that determines seed color and we call them *alleles* of each other, or *allelomorphs* (*morph* = form; *allelon* = of each other).

Figure 4.1 shows how easily Mendel's model explains his results. His purebred yellow-seed strain must carry two *Y* factors (*YY*), and his purebred green-seed strain must carry two *y* factors (*yy*); since both factors in these strains are the same, we say they are *homozygous* or that the plants are *homozygotes.* Each of the original plants contributes one factor for seed color, so the F_1 plants are all *Yy*; since they carry different factors, these plants are *heterozygous* – they are *heterozygotes.* When these heterozygotes reproduce and are crossed with each other, each one produces two kinds of gametes, half carrying *Y* and half *y*. These gametes combine at random to produce one of four combinations: *YY*, *Yy*, *yY*, or *yy*. Only the latter, having two recessive *y* factors, is green; the other three are yellow, thus explaining the 3:1 ratio that Mendel observed.

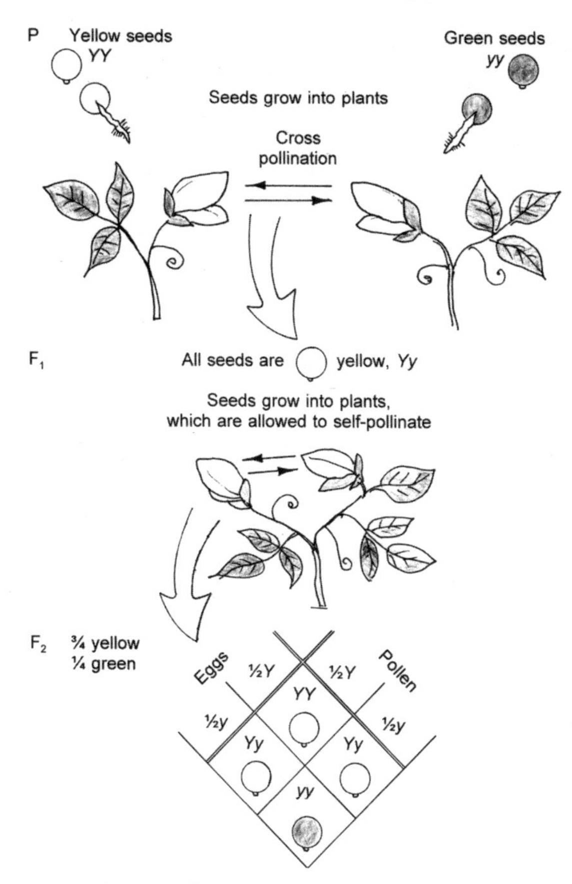

Fig. 4.1. *An explanation of Mendel's results. Each individual has two copies of the gene for seed color, but it passes on only one copy in each of its gametes. The* Y *gene is dominant to* y*, so all of the F_1 plants, being* Yy*, are yellow. When these events are repeated for the next generation, they produce four combinations, of which three are yellow and one is green.*

pedigrees

In addition to using the large numbers that come from mating plants and animals at random, it is instructive to examine the patterns of inheritance in humans and domestic animals, which are shown as a *pedigree* or family tree:

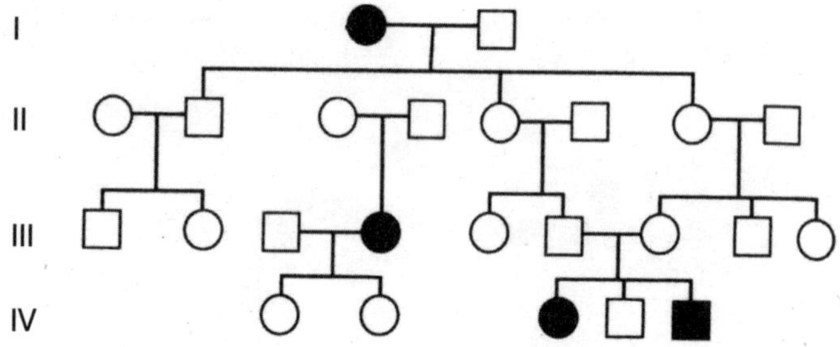

Females are designated by circles and males by squares; a diamond indicates an individual whose sex is unknown (such as an obscure relative who died in childhood). A horizontal line connecting a male and a female indicates a mating. The children from each mating – referred to as *siblings* or *sibs* – are shown as branches stemming from one horizontal line, and their order of birth is read from left to right. Notice that the original pair with which a pedigree begins is designated generation I, and their descendants are generations II, III, and so forth. Twins are indicated by lines leading to the same point in the parental line and identical twins by a line connecting them. An abortion or miscarriage is represented by •. A pedigree showing no male parent means that the father of the children is unknown.

When a specific character is being followed, individuals having the trait are designated by a mark in the symbol, such as ● or ⊗. A dot in the symbol designates an individual who does not exhibit the trait but who carries the factor or factors determining it.

Now we can use human pedigrees to further illustrate the principles Mendel established with plant experiments. A familiar hereditary trait is *albinism*, the absence of pigments normally found in skin, hair, and eyes. Albinos appear occasionally in all racial groups, with a frequency of about one in twenty thousand people among north American Caucasians. Although albinism is generally rare among native North Americans, up to one in every 200–300 Hopi and Zuni people of the southwestern United States is an albino, for among these peoples albinos are regarded as special and are encouraged to reproduce.

When two albinos marry and have children, all their children are albinos. One such family had this pedigree:

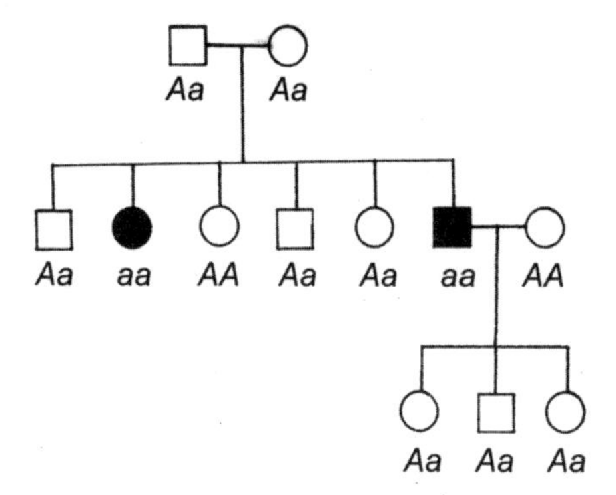

With rare exceptions, albinos marry normal, pigmented individuals, and generally (where there is no history of albinism in the pedigree of the normal mate), their offspring are all normal:

I	Albino × Normal
	↓
II	All normal

From a series of matings like this, we can derive some important lessons and uncover the basic laws that Mendel discovered.

mating 1

I	Albino × Normal
	↓
II	Normal × Normal
	↓
III	All normal

This situation shows that one trait does not *dilute* another. All the individuals are normally pigmented; we see no albinos with slightly darker skin or normal people with slightly lighter skin. Indeed the albino trait seems to have disappeared. But a different mating shows that it has not.

mating 2

Suppose we mate the offspring of two different albino × normal crosses:

I	Albino × Normal		Albino × Normal
	↓		↓
II	Normal	×	Normal
		↓	
III		Most normal, a few albino	

This shows that the trait has not disappeared; it has only been hidden somehow, and it can show up in later generations. Albino × albino matings, which produce only albinos, show that albinos do not even have the *potential* to produce anything else. However, some offspring of albino × normal matings have the potential to produce albinos, even though they appear normal. Thus, *an organism can carry a genetic potential that it does not exhibit.*

Mendel's simple model easily explains these results. First, we assume that the difference between albinism and normal skin coloration is controlled by a single gene. Second, as in Mendel's pea analysis, we assume that every individual has two copies of each gene, one obtained from each parent. Third, we say that the gene determining skin pigmentation has two alleles: a normal, dominant allele denoted by *A* and a recessive allele for albinism denoted by *a*. A specification of the genes that an individual carries is called the *genotype*, so a normal person (who has no albino ancestors) must have the genotype *AA* and an albino must be *aa*. These are both the genotypes of homozygotes. Normal people can pass along only *A* genes and albinos can pass along only *a* genes, so all the offspring of a normal × albino mating, who receive one gene from each parent, must be heterozygotes with the genotype *Aa*. These *Aa* heterozygotes look just like homozygotes with the genotype *AA*; in genetic terms, they have the same expressed characters, called their *phenotype*. The standard terminology for the three kinds of individuals is: homozygous dominants, *AA*; homozygous recessives, *aa*; and heterozygotes, *Aa*.

In a mating between two *Aa* heterozygotes, as in Mating 2, most of the offspring are normal, but a few are albinos. Taking a lesson from Mendel, however, we should count individuals, rather than saying "most" and "a few." Although it isn't as easy to study large numbers of people as it was for Mendel to get large numbers of peas, the summed results from many matings show the same simple 3:1 ratio – three normal people to one albino – that Mendel found with his peas. We understand it by means of the same simple model. During the formation of gametes, the two genes in each pair *segregate* from each other, so each gamete contains only one of them. This point, known as Mendel's *law of segregation*, means that each of the *Aa* father's sperm carries only one allele, so half his sperm carry *A* and half carry *a*. Similarly, half of the *Aa* mother's eggs carry *A* and half carry *a*.

Since fertilization occurs at random, there are four possible combinations:

A egg and *A* sperm: genotype *AA*
A egg and *a* sperm: genotype *Aa*
a egg and *A* sperm: genotype *Aa*
a egg and *a* sperm: genotype *aa*

Since the first three combinations will be normal people and the fourth will be an albino, the model explains the observed ratio of three to one.

The model explains why only normally pigmented people appear in the descendants from the mating of a normal person with an albino, provided no one marries another albino. The offspring of the mating, in generation II, are heterozygous, and many people in later generations will also be heterozygous. They all pass on the *a* gene in half their gametes, but if these *a* genes combine exclusively with *A* genes from *AA* individuals, no *aa* people can ever appear.

The one type of mating that we haven't considered yet can test our model further.

mating 3

I Albino × Normal
↓
II Normal × Albino
↓
III ?

What can we predict? Well, we know from previous analysis that the normal person in generation II (let's call him a male) must have the genotype *Aa*. Therefore, half his sperm carry *A* and half *a*. The albino has the genotype *aa*, so all her eggs carry *a*. Thus there are only two possibilities:

a egg and *A* sperm: genotype *Aa*
a egg and *a* sperm: genotype *aa*

So half the offspring should be *Aa*, with normal pigmentation, and half should be *aa*, albinos. And this is what we find.

Note that the simple model we have developed makes only a few assumptions, which are all reasonable and justified:

- that every individual has two copies of each gene;
- that some alleles may be recessive to others;

- that genes segregate from each other in the formation of gametes;
- that gametes combine at random to make new zygotes.

Yet this model gives us a powerful tool for thinking about heredity, and its ability to explain what we observe indicates that it must be a good model, one that agrees well with reality.

another example: tasters and non-tasters

Now let's try another example for practice. A chemical called phenylthiocarbamide (PTC) tastes very bitter to some people (tasters) but is tasteless to others (non-tasters). About seventy percent of North American Caucasians are tasters. The taster and non-taster phenotypes are hereditary, the taster allele being dominant. Let *T* represent the taster allele and *t* the non-taster allele. It is no surprise that a cross of homozygous tasters (*TT* × *TT*) yields only taster offspring; and a cross of homozygous non-tasters (*tt* × *tt*) yields only non-tasters. But now consider a cross of these two homozygotes:

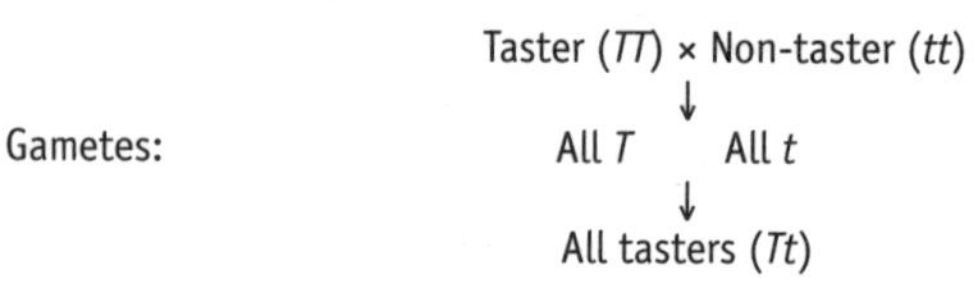

Some crosses involving heterozygotes yield these results:

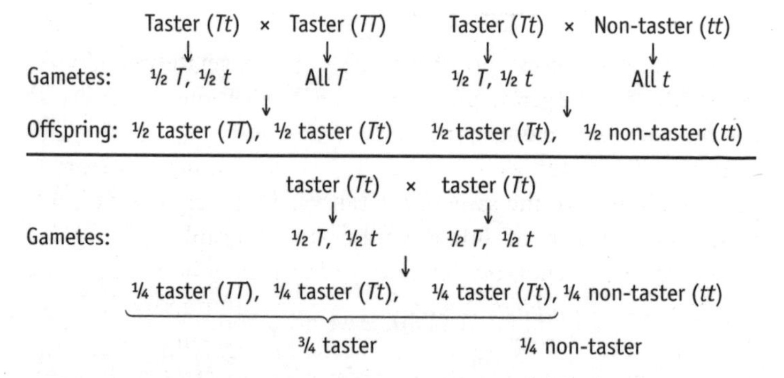

The most illustrative cross is *Tt* × *Tt*. When you understand the phenotypic ratio of 3:1 in their offspring, you understand Mendel's law of segregation. One simple way to work out the results of a cross is to use a *Punnett square*, a grid in which the headings of the rows and columns show the gametes produced by each sex. Thus, in the *Tt* × *Tt* cross, the Punnett square is drawn in the following way:

Sperm \ Eggs	½ *T*	½ *t*
½ *T*		
½ *t*		

Each square represents the union of the two gametes:

Sperm \ Eggs	½ *T*	½ *t*
½ *T*	¼ *TT*	¼ *tT*
½ *t*	¼ *Tt*	¼ *tt*

A Punnett square shows more graphically how we get the phenotypic ratio of 3:1 and the genotypic ratio of one homozygous dominant to two heterozygotes to one homozygous recessive.

In the examples we've discussed so far, homozygotes and heterozygotes have the same dominant phenotype. However, the Mendelian pattern of inheritance is more graphic when the heterozygote is phenotypically different from either homozygote, as in cases of *incomplete dominance*, a situation in which the phenotype of heterozygotes is intermediate between those of the homozygotes. In some plants, for instance, a cross of a red-flowered with a white-flowered strain produces hybrid progeny with pink flowers; then mating the pink-flowered heterozygotes produces a Mendelian ratio of one red to two pink to one white. In this situation the 1:2:1 *genotypic* ratio is directly reflected in the 1:2:1 *phenotypic* ratio.

blood types

Some excellent genetic lessons come from the study of blood types. Blood is categorized into types according to its interactions with the body's immune system, which defends against invasions by foreign bodies such as bacteria, viruses, and fungi. This system, which is developed best in birds and mammals, depends on the distinctiveness of molecular shapes, especially the shapes of proteins; all cells have many distinctive proteins and other molecules on their surfaces, and each kind of organism has a distinctive set of them. They give our tissues individuality, which makes it hard to successfully transplant kidneys and other organs from anyone but our closest relatives. Foreign molecules get into our bodies accidentally, as through a wound or through medical practice – for instance, by an injection of vaccine that is intended to develop immunity against some disease. The immune system recognizes these molecules as foreign, so they are now, by definition, *antigens*; an antigen is any substance that induces a response from the immune system. The presence of a specific antigen causes certain cells in the immune system (lymphocytes) to make specific proteins called *antibodies* that recognize and counteract that antigen. Each antibody has a shape that is complementary to the antigen that induced it, so the two molecules bind to each other:

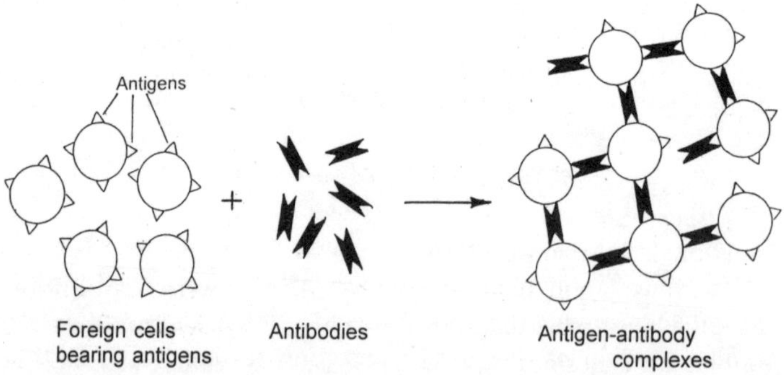

Antigens

Foreign cells bearing antigens

Antibodies

Antigen-antibody complexes

This binding may inactivate the antigen, or the molecules may form large molecular complexes that are detected and removed by certain white blood cells (neutrophils and macrophages), which are rather like garbage collectors. For instance, bacteria that get into a cut are recognized as foreign because of their distinctive surface proteins and polysaccharides which act as antigens. The body makes antibodies against them, which bind to the bacteria and clump them together so they can be eliminated.

In 1927, Karl Landsteiner and P. Levine found that injecting human blood cells into rabbits induces them to develop antibodies against antigens on the human cells. By virtue of the antibodies made to the blood of different people, Landsteiner and Levine identified two types of antigens which they called M and N. Cells from a person with type M blood induce the rabbits to make anti-M antibodies; cells from a person with type N blood induce the rabbits to make anti-N antibodies. Every person has blood of type M, type N, or type MN – a mixture of M and N antigens.

Matings of people with known MN blood phenotypes give these results:

M × M: all M children
N × N: all N children
M × N: all MN children

This implies that the M and N people are homozygous for different alleles of a single gene and that MN people are heterozygotes in which *both* alleles are expressed. The gene for this character is called *L*, in honor of Landsteiner; its two alleles are designated L^M and L^N, which determine the M and N antigens, respectively. These alleles are *codominant*, which means that in $L^M L^N$ heterozygotes the alleles are equally expressed. This simple model explains the three matings above. Furthermore, in a mating between heterozygotes, each parent makes half L^M and half L^N gametes, and they should combine as follows:

$L^M L^M$	$L^M L^N$	$L^N L^M$	$L^N L^N$
1/4 M	1/2 MN		1/4 N

This, in fact, is what we observe.

Much earlier, in 1900, Landsteiner had discovered that humans have the blood types that we now commonly think about in relation to blood transfusions: A, B, AB, and O. These differences show up because of the occasional clumping of red cells during attempted blood transfusions. Transfusions between people of identical type are always safe, but other transfusions promote a severe adverse reaction because the transfused cells of one type clump when they enter the blood of a different type. This adverse reaction occurs because our red blood cells may have the A and B antigens on their surfaces and because people normally have antibodies against the A and B antigens that they do *not* have. Type A people have only A antigens on their blood cells and have antibodies in their serum against type B blood cells; the opposite is true for type B people. The red blood cells of type O people have neither A nor B antigens, so these cells do not react with anyone's serum; type O people are universal donors whose blood is accepted by everyone. Type AB people, having both kinds of antigens on their blood cells, have neither anti-A nor anti-B antibodies in their serum, so they are universal acceptors who can accept blood of any type.

The genetics of the A–B–O system is actually quite simple and instructive. These types are determined by *three* alleles of a single gene denoted by I, a simple example of *multiple alleles* – three in this case, rather than the two alleles of previous examples. The allele I^A determines type A antigens, I^B determines type B antigens, and i specifies no antigen at all of this kind. Both I^A and I^B are dominant to i – a fact reflected in upper- and lower-case notation. A type O person has the genotype ii. A type A person could be either $I^A I^A$ or $I^A i$, and a type B person could be either $I^B I^B$ or $I^B i$. However, I^A and I^B are codominant with each other, just as L^M and L^N are, so a person of genotype $I^A I^B$ has blood type AB.

You should now be able to predict what kinds of children can result from various matings, with respect to blood type. For instance, take two type A parents. Their genotypes may be either $I^A I^A$ or $I^A i$; since we don't know which, we denote such a genotype by I^A –, where the dash means that there are at least two possibilities for the second allele. These people can obviously pass along I^A genes, and we expect their children to be type A. If only one of them were heterozygous, *all* their children would be type A. However, if they were both heterozygous, a quarter of their offspring, on average, would receive the i allele from both parents and would be type O.

Now test yourself by predicting the possible offspring from the following matings. The answers are at the end of the chapter.

1. $I^A I^B \times I^A i$ 2. $ii \times I^A i$
3. $I^A I^A \times I^B i$ 4. $ii \times I^A I^B$

Simply stated, Mendel's model is important because we can use it to show that many pairs of discrete phenotypic differences in plants and animals are inherited as allelic differences of a single gene. In humans, Mendel's concept gives us a powerful law that we can use to predict the hereditary patterns of normal variants such as blood groups and abnormal conditions such as certain genetic diseases. Many diseases in humans are known to be controlled by alleles of single genes. Some examples are the disease phenylketonuria (PKU), caused by a recessive allele we might call *p*; Tay–Sach's disease, caused by a recessive allele *t* of a different gene; and Huntington's disease, caused by a dominant allele *H* of still another gene. These alleles are inherited precisely in the pattern that Mendel would have predicted. Note, however, that some diseases have a far more complex hereditary basis, and of course some diseases are not hereditary at all. Only the analysis of appropriate pedigrees can determine which situation applies. We will look more closely at the basis of single-gene diseases in later chapters.

multiple alleles and dominance

Phenomena such as incomplete dominance and codominance show that interactions between alleles of a single gene can be complicated. The A–B–O blood groups are determined by three alleles of one gene, and other genes have even more alleles, sometimes related in complicated ways. They show that dominance and recessiveness is a relationship between alleles, not an absolute characteristic of one allele. The fur color of rabbits, for instance, is determined by a gene *c* with four alleles: c^+, c^{ch}, c^h, and c^a. The c^+ allele is dominant to the other three, so a c^+c^+ homozygote or any heterozygote with one c^+ allele has wild-type dark gray fur, sometimes called agouti. A $c^{ch}c^{ch}$ homozygote has tan chinchilla fur, but the c^{ch} allele is incompletely dominant to both c^h and c^a, so a $c^{ch}c^h$ or $c^{ch}c^a$ rabbit has light gray fur. The c^h allele is then dominant to c^a, so a c^hc^h or c^hc^a rabbit has the Himalayan pattern, mostly white but with black nose, ears, feet, and tail. Finally, c^ac^a produces a pure white albino rabbit.

test crosses

An organism with a dominant phenotype for some character could be homozygous or heterozygous – *AA* or *Aa*, let us say. It may be important to know its genotype; for instance, if you anticipate using it for breeding, you may want to be sure it is homozygous for a desirable trait. You can determine its genotype with a *test cross*. You cross the unknown individual with a homozygous recessive, *aa*, known in such a situation as a *tester*. An *AA* parent can produce only *A* gametes, so all the offspring will be heterozygotes with the dominant phenotype. However, a heterozygous parent will produce half *a* gametes, and therefore half of the offspring should be homozygous recessives, *aa*. In fact, if any *aa* progeny are found the parent must have been heterozygous. This is a simple, informative test.

The same reasoning can be used with humans, although human matings cannot be called test crosses. If we consider a family in which one parent has type A blood and the other has type O, their offspring can reveal whether the type A parent is homozygous or heterozygous. The appearance of even a single type O child indicates heterozygosity. However, if the couple has only a few children and all are type A, we cannot conclude that the parent is homozygous, because with so few tosses of the genetic dice it is quite possible that the dominant allele from a heterozygous parent just happened by chance to be inherited in all cases.

probability

Mendel's law of segregation allows us to predict the probability of inheriting certain traits. Mendel initiated quantitative methods in genetics by counting large numbers of individuals and treating his numbers in a simple statistical way; without this approach, genetics would never have gotten off the ground. We have to introduce some simple ideas about probability to go any further.

Probability is a hard concept for many people to understand. Experience with the parents of genetically defective children (suffering, for example, from phenylketonuria, cystic fibrosis, or Down's syndrome) shows that few understand the biological basis of the disease or why it has a certain probability of occurring, even when they have been counselled by geneticists. Even physicians tend to shy away from the subject of probability, and the medical geneticist Judith Hall has noted that "The medical profession does not like the idea of probability; they like the idea of absolutism. They like absolutism because it is more appealing to their patients." Yet we must run our lives every day by estimating probabilities. Sometimes it is an intuitive feeling for the "odds" that motivates us, sometimes a rational judgement based on the analysis of hard figures by actuaries. Genetics, too, needs this kind of firm base. We often want to determine the chances that we have the same allele as some relative. This requires some simple math.

We all know that the chances of flipping either heads or tails with an honest coin are even, or 50:50. This is the same as saying that the probability of flipping a head is ½, or 0.5. Similarly, there are equal probabilities that an unloaded die will land with any one of the six sides upward, so the probability of getting any number is ⅙. This is rather obvious.

But how do we deal with the probability of two or more events together, such as the chances of getting two heads with two coins flipped simultaneously. Suppose we do this many times. Half the time, the first coin should come up heads; but half of *that* time the second coin will come up heads and the other half it will come up tails. Similarly, the first coin will come up tails half of the time; again, in half of those times the second coin will come up heads and the other half it will come up tails. So there are four equally probable cases, each with a probability of ¼:

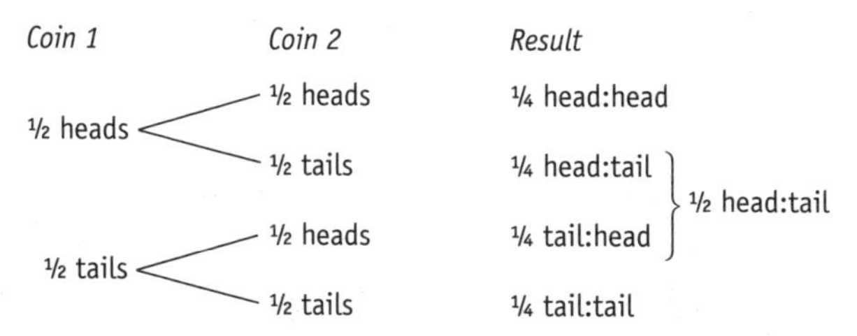

This leads us to one law of probability. Given two independent events, where the occurrence of one has no influence on the other, the probability that they will occur together is the *product* of their individual probabilities. When applied to the coins, this product rule means that the probability of getting heads on both coins is the chance for the first coin (½) times the chance for the second coin (½), which equals ¼. The same is true for the other three cases.

Now this is exactly the kind of situation we have encountered already in determining the ways a sperm and an egg can be combined. In a cross between two *Tt* heterozygotes, for instance, each one makes ½ *T* gametes and ½ *t*. The Punnett square then gives us the four possible combinations, each with a probability of ½ × ½ = ¼, thus accounting for the standard ratios of 1:2:1 or 3:1.

Now suppose your maternal grandfather was an *Aa* heterozygote. What is the probability that you've inherited his *a* allele? Well, the chance that your mother received it from him is ½ and the chance that you received it from her is ½, so the probability is ¼. This is just like flipping a coin twice in a row. The probability is exactly the same that you have received any specific allele from any gene pair in any grandparent.

When would two events *not* be independent? Suppose for some reason that an *A* egg tends to attract *A*-bearing sperm and an *a* egg tends to attract *a*-bearing sperm. Then in a mating of an *Aa* male and an *Aa* female, both producing equal numbers of *A* and *a* eggs, the probabilities of getting the various offspring would be skewed from what we usually expect, since fertilization would not occur at random.

two or more genes

An important application of these concepts of probability is in predicting the results of crosses where we follow two or more genes simultaneously. Mendel performed such experiments, in which he followed characters such as seed color and seed shape together, and they led him to a second important principle, the *law of independent assortment*: the alleles of two genes assort independently when gametes are being formed. In Chapter 5, we will show how Mendel's laws follow from the way chromosomes move during cell division. Mendel knew nothing about chromosomes and their behavior in reproduction, and he derived the law of independent assortment strictly from patterns of inheritance, but he was lucky. We shall see later that many genes occur together on the same chromosome and therefore tend to be inherited together, so these are *not* independent events. However, it just happened that the genes for all the characters that Mendel studied are on different chromosomes (or are so far apart on one chromosome that they are essentially independent). We can illustrate Mendel's principle with human heredity and use the rules of probability to predict what combinations of traits will occur.

Consider a couple who are simultaneously heterozygous for two genes: they are both *Tt*, for the taster gene, and both *Bb* for a second gene that determines brown eyes (*B*–) or blue eyes (*bb*). Now consider how they can make gametes. By Mendel's first law, the *T* and *t* alleles will segregate from each other, so half the gametes will carry *T* and half will carry *t*. By his second law, the *B* and *b* alleles will segregate from each other independently, so half of these cells will carry *B* and half will carry *b*. Therefore, just as in flipping coins, there are four combinations – *TB, Tb, tB,* and *tb* – and the probability of each of them is ¼.

Each parent produces gametes in exactly the same way. Then how can they be combined? Since any of the four types of sperm can be combined with any of the four types of eggs, there are sixteen combinations, and the probability of any one of them is $\frac{1}{4} \times \frac{1}{4} = \frac{1}{16}$. One way to see these combinations is to draw a larger Punnett square (Figure 4.2). To determine the phenotype of each combination, we first find squares with at least one *T* allele and at least one *B* allele, so they will be brown-eyed tasters; there are nine such squares. Three squares have at least one *T* allele plus the *bb* combination, so these are blue-eyed tasters. Another three squares have the combination *tt* with at least one *B* allele, so these are brown-eyed non-tasters. Finally, one square has blue-eyed non–tasters, with the genotype *tt bb*. This classical ratio of 9:3:3:1 is the result of a mating between two heterozygotes for two genes that are inherited independently of each other.

However, by thinking about the probability of getting each phenotype, we can predict this result more easily without drawing the square – a rather tiresome exercise in which it is easy to make a mistake. Since the two genes are inherited independently, we *think* about them independently. If we were simply mating *Tt* heterozygotes, we would expect to get a ratio of ¾ tasters to ¼ non-tasters. Similarly, in mating *Bb* heterozygotes we would expect to get

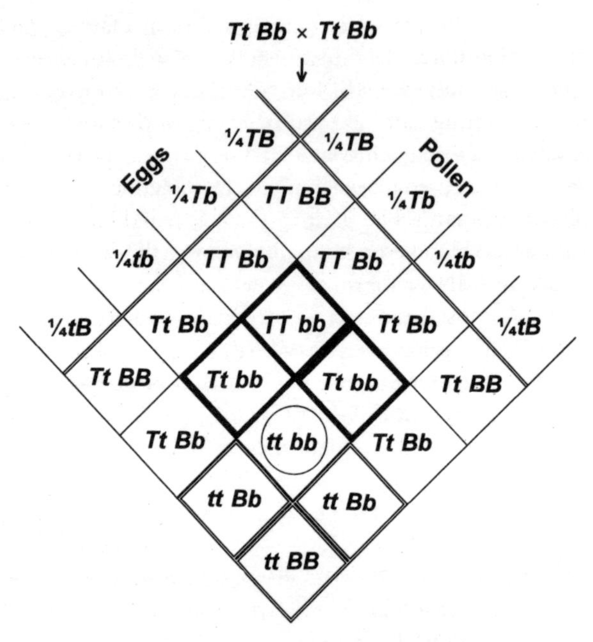

Fig. 4.2. *A Punnett square shows the possible offspring of a cross between two individuals who are heterozygous for two independent characters. The phenotypes fall into four classes in the ratio 9:3:3:1.*

¾ brown-eyed and ¼ blue-eyed individuals. These are the elementary probabilities. Now, as before, we simply multiply them:

¾ [taster] × ¾ [brown] = 9/16
¾ [taster] × ¼ [blue] = 3/16
¾ [non-taster] × ¼ [brown] = 3/16
¼ [non-taster] × ¼ [blue] = 1/16

By thinking in this way, we could easily predict the results for three, four, or any number of genes simultaneously, and we can easily apply the method to crosses in which the probabilities are different.

mendel's first law and disputed paternity

Through elementary Mendelian reasoning, modern geneticists can use pedigrees to determine the mode of inheritance of a character and to determine the probability that someone in a family will have the trait. Furthermore, Mendel's principles can help to resolve cases of disputed paternity. We shall see later that a new method called DNA fingerprinting can establish paternity with some certainty. Classical genetics can seldom prove that someone *is* the father of the child, but it can be used to rule him out unambiguously. As a simple example, if a woman is blood type M ($L^M L^M$), and her husband or boyfriend is also blood type M, he could *not* be the father of a child with blood type MN ($L^M L^N$).

One of the most famous cases of disputed paternity involving genetic evidence was Joan Berry's claim that Charlie Chaplin was the father of her daughter. In the case, stretching from 1943 to 1945, Chaplin was vilified by the Hearst press on moral grounds. Headlines of the type on the front page of the *New York Times* – "Chaplin, 6 others indicted for plotting against girl" (11 February 1944) – preceded his conviction and loss of appeal. The genetic facts were that Chaplin had blood type O, Joan Berry (the mother) was type A, and the daughter, Carol Ann Berry, was type B. Since the child did not have type A, her mother must have been heterozygous, $I^A i$, and the child must have received the i allele from her mother. The I^B allele must have come from the father, who could not have been Chaplin. Nevertheless, Chaplin was convicted in California, and the conviction was subsequently upheld by the District Court of Appeal, which justified its decision by stating

> The evidence concerning the blood test is expert opinion because the conclusions reached by the examiner are based upon medical research, and involve questions of chemistry and biology with which a layman is entirely unfamiliar, but that such tests and the evidence thereof is not conclusive because not so declared by the code (code Civ. Proc. sec. 1978); ... when scientific testimony and evidence as to the facts conflict, the jury or the trial court must determine the relative weight of the evidence.[2]

We can only hope that the courts have progressed since then, but the murder trial of the football star O.J. Simpson suggests that courts still sometimes allow less compelling evidence to outweigh genetic evidence.

Another interesting paternity dispute involved an instance of superfecundation, where a woman became pregnant after having sexual intercourse with two different men in the same night. She subsequently gave birth to twins, who were unequivocally shown to have had two different fathers. The observations were:

	Phenotype	*Genotype*
Mother	0	ii
Twin 1	B	$I^B i$
Twin 2	A	$I^A i$
Male 1	A	$I^A -$
Male 2	B	$I^B -$

You can see that twin 1 had to be the child of male 2, and twin 2 was the child of male 1.

Genetic analysis is a powerful predictive tool. When a character is known to be determined by a pair of alleles, a genetic counsellor can assign probabilities for the occurrence of a particular phenotype. For instance, we can understand how a couple can have a child with cystic fibrosis, caused by a recessive allele. Since the parents are themselves normal, they each carry a normal dominant allele. Since they have a homozygous recessive baby, however, they must each carry a recessive allele (that is, they are heterozygous). Thus, if they choose to have another child, they will have a ¼, or twenty-five percent, chance of giving birth to a second child with the disease.

Similarly, plant and animal breeders who want to develop new lines that breed true for a given character can apply Mendel's laws to estimate the number of crosses that are required. This provides a very precise method on which modern agricultural breeding programs can be built.

answers to blood types questions

1. $I^A I^B \times I^A i$: ½ A, ¼ B, ¼ AB
2. $ii \times I^A i$: ½ A, ½ 0
3. $I^A I^A \times I^B i$: ½ A, ½ AB
4. $ii \times I^A I^B$: ½ A, ½ B

chapter five

chromosomes, reproduction, and sex

Mendel's discovery of the basic principles of inheritance came at about the same time that other biologists were learning about the structures of cells, so when his laws were rediscovered in 1900, they could be clearly understood by relating his "factors" to real structures, to chromosomes. Mendel's "factors" are what we now call genes, and each chromosome carries many genes. Connecting abstract factors to real objects was a triumph of early genetics, and it was closely connected to the development of ideas about the determination of sex.

cells and reproduction

After the cell theory of Schleiden and Schwann had become generally accepted, the pathologist Rudolph Virchow added an important general point: not only are all organisms made of cells, he proposed, but every cell comes from a pre-existent cell. We have shown that growth generally occurs as a cell takes in nutrients, makes more of its own substance until it is about twice its original size, and then divides in two. Cell division is also the simplest kind of reproduction. It is the most common way that unicellular organisms such as protozoa and yeast reproduce. The cells of multicellular bodies such as ours also reproduce in this way. Although some of our cells grow into a specialized form and never divide again, many other cells (in the skin, liver, lining of the intestinal tract, and elsewhere) are continually being lost and replaced by new growth. As you read this, in fact, your body is destroying old red blood cells at the rate of about two million per second and producing an equal number through the growth and division of cells in your bone-marrow.

So all organisms, being made of cells, come from pre-existing organisms. The idea of spontaneous generation of organisms from some inorganic source was laid to rest in the nineteenth century, principally through the experiments of Louis Pasteur. Then how does sexual reproduction fit into the picture? Though the answer seems obvious to us today, centuries of painstaking observation and testing were needed to arrive at our present state of knowledge. As we showed in Chapter 2, microscopes revealed spermatozoa in semen and, eventually, eggs released by the ovaries (Figure 5.1).

For a long time, scientists believed in *preformation*, the idea that the reproductive cells contain completely formed replicas of the

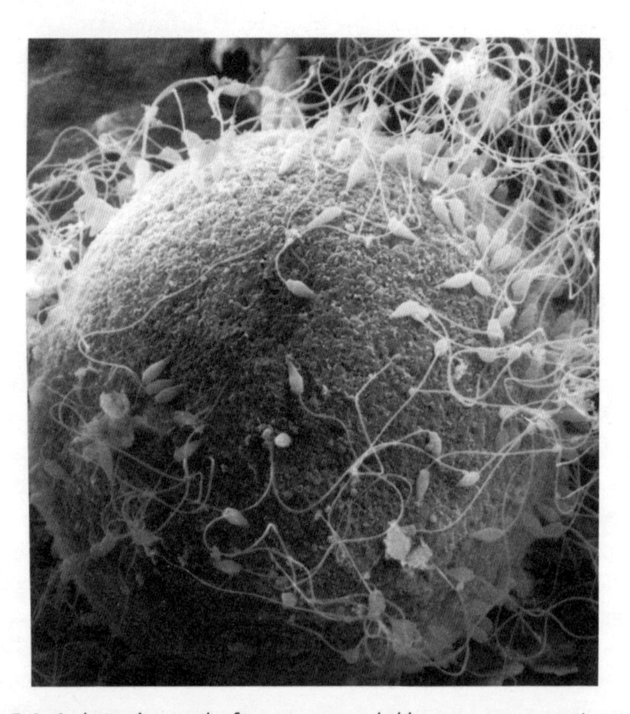

Fig. 5.1. *A photomicrograph of an egg surrounded by numerous spermatozoa at the moment of fertilization.*

adult, which simply grow larger as the embryo develops. But in 1759 Caspar Friedrich Wolff observed developing chicken embryos with a microscope and showed that an embryo develops through *epigenesis.* This means that its body parts form progressively as it grows. Development begins when a spermatozoon fertilizes an egg to form a *zygote.* The organs and parts of the body are made of *somatic* cells (such as muscle, bone, and liver cells); gametes, in contrast, are derived from special *germ* cells that reside only in the ovaries and testes. Thus, life extends back to previous cells in an unbroken chain from generation to generation, ultimately leading back to common ancestors in the distant past.

mitosis and the cell cycle

A single cell grows and divides into two by going through a *cell cycle.* The whole function of the cell cycle is to make two or more identical cells that can repeat the process, so each cell must have its own copy of the genome. Since the genome is carried mostly in the chromosomes that reside in the nucleus, the cell must first make an extra set of these chromosomes. The chromosomes are duplicated during the *S phase* of the cycle (S for DNA *synthesis*), the period when all the DNA is replicated, so the genome is doubled. The chromosomes then divide into identical sets during a process of *mitosis* (M), a nuclear division that produces two nuclei with identical sets of chromosomes:

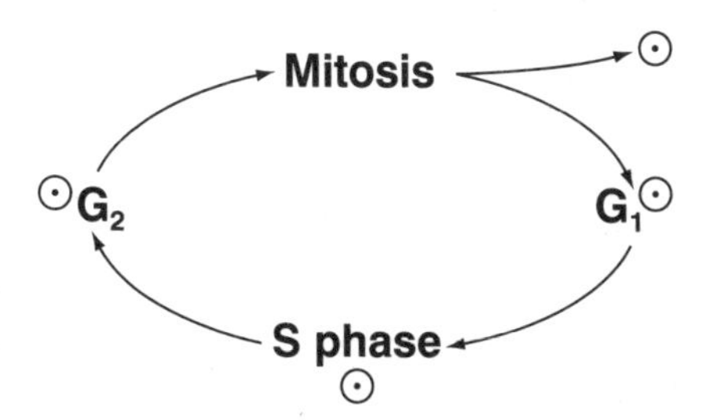

Thus, the cell cycle centers on two major events: a time when the chromosomes are duplicated and a time when the cell divides so that each daughter cell gets one copy of each chromosome.[1]

Between M and S is a gap called G_1, a relatively long period of cell growth. Between S and M is a second gap, G_2, when the cell is preparing for mitosis. The whole time G_1 plus S plus G_2 is sometimes called *interphase*, an old term left over from the days when people only knew about the events of mitosis and thought they were the important part of the cycle. Mitosis is divided into several stages, which are all phases of a continuous dynamic process. We shall explain the process of DNA replication in more detail in Chapter 7, where we discuss DNA structure. For now, the important point is that after passing through the S phase, all the chromosomes in a cell have been duplicated and are ready to be divided into identical sets in two daughter cells.

During interphase, we can see little happening inside the nucleus. As the cell enters the first stage of mitosis, *prophase* (Figure 5.2), the nuclear envelope forming the boundary of the nucleus disintegrates and we can see that it contains several distinct, ropelike chromosomes. At this stage, each chromosome is doubled and consists of two side-by-side "ropes" called *chromatids*. They are *sister* chromatids – two identical copies that were made when the DNA was replicated during the S phase; the sister chromatids are joined at the *centromere* of each chromosome. At one side of the nucleus, in animal cells, a pair of tiny bodies called *centrioles* start to move toward opposite sides of the cell, where they will form *division poles*. A structure called the *spindle* forms between the centrioles; it is made of many fibers (actually, very fine microtubules made of the protein

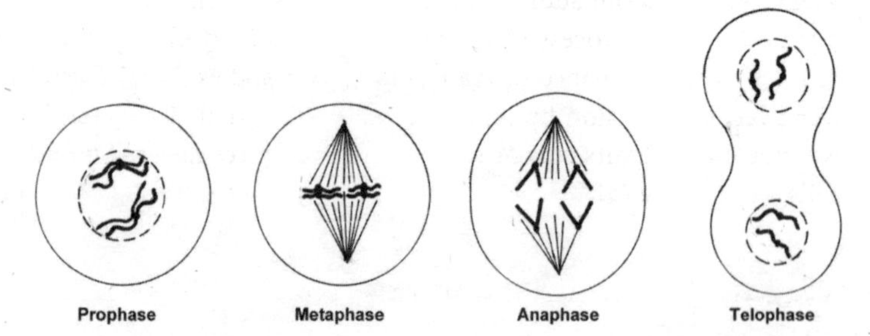

Fig. 5.2. *Cells in the stages of mitosis. (a) Prophase, when the chromosomes are becoming visible and the nuclear envelope is breaking down. (b) Metaphase, when the chromosomes line up along the middle of the cell. (c) Anaphase, when the two chromatids of each chromosome separate and move toward opposite poles. (d) Telophase, when new nuclear envelopes form and the chromosomes become indistinct again.*

tubulin) that will eventually separate the chromosomes into two nuclei. Some of these fibers run from the centriole to the centromere of a chromosome, and they are in a position to pull the chromosomes along.

The chromosomes are pulled back and forth across the cell as spindle fibers attach to them, but soon they settle down midway between the poles (in the *equatorial plate*). The cell has now entered *metaphase*, the second phase of mitosis. Suddenly, as if at a command, the sister chromatids of each chromosome separate and begin to move toward opposite poles, drawn along by shortening of the spindle fibers, while other fibers elongate to push the poles farther apart. These events mark *anaphase*, and later in this stage the cell begins to divide in two by pinching in around the middle.

Finally, in *telophase*, the chromosomes reach the poles, the spindle apparatus disperses, and new nuclear envelopes start to reform around the chromosomes as they lose their distinctness. The cell then completes its division into two daughter cells.

The result of this elaborate process is the exact distribution of two identical sets of chromosomes to the daughter cells, so they both have the same chromosome complement as the original mother cell. Thus, mitosis conserves the chromosome content of a cell line from cell generation to cell generation. Furthermore, these visible features of mitosis reflect a complex sequence of exquisitely coordinated molecular processes that are repeated over and over to transform a single zygote into an adult comprising trillions of cells. Not only is cell division the process of growth in animals and plants, but it ensures the maintenance of the health of our bodies. Every day of our lives, cell division by mitosis replaces skin that is constantly wearing away, repairs cuts or wounds, and produces new red blood cells.

karyotypes

We can use our knowledge of mitosis to get a better look at the chromosomes that are being moved around in this process. We put a small blood sample into a tube of nutrient medium in which white blood cells can grow. After giving them a few days to proliferate, we treat them with the drug *colchicine*, which disrupts the spindle apparatus and stops all dividing cells in metaphase, when their chromosomes are maximally condensed and visible. T.C. Hsu found that if cells are put into a solution with a lower salt concentration than living cells have, they take up water and swell so that their chromosomes uncurl and spread out for easier observation. These cells are spread on a microscope slide to display their chromosomes clearly and photographed (Figure 5.3(a)). We can then see that the chromosomes have different lengths and shapes; some are long and some short, and their centromeres are located in different positions. Also, every species has a definite, characteristic number of chromosomes – forty-six in humans. In humans and most other animals, the chromosomes can be put together in pairs. The forty-six human chromosomes form twenty-three pairs (Figure 5.3(b)), and a photograph showing them neatly lined up in order is called a *karyotype*; it can be invaluable for diagnosing certain genetic disorders.

Two chromosomes that look identical are called *homologs* – they are said to be *homologous.* We start numbering the human chromosomes with the longest pair, and when we get to the shortest ones, we find an interesting difference between males and females. Whereas females have twenty-three pairs, males have twenty-two pairs plus two odd ones, one of them extremely short. This very short one is a *Y chromosome*, and the longer odd one is an *X chromosome.* Then we see that pair twenty-three in females is actually two X chromosomes. It is clear that the X and Y chromosomes are related to sex. The other twenty-two pairs of homologs, which all people have, are *autosomes.* It should be obvious that people have two of each chromosome because everyone has two parents. Everyone is formed by the union of a sperm and an egg; each gamete carries just twenty-three chromosomes – one of each kind – and the resulting zygote has two of each kind.

This leads us, conveniently, to the whole matter of sexual reproduction.

meiosis

Recall that mitosis occurs in ordinary somatic cells and maintains exactly the same set of chromosomes in all of them. But if eggs and sperm were produced by mitosis, a zygote would have twice the

(a)

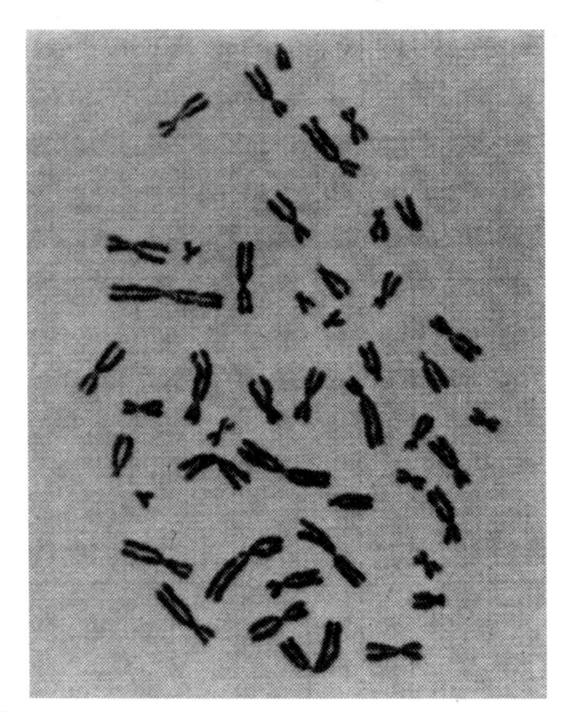

(b)

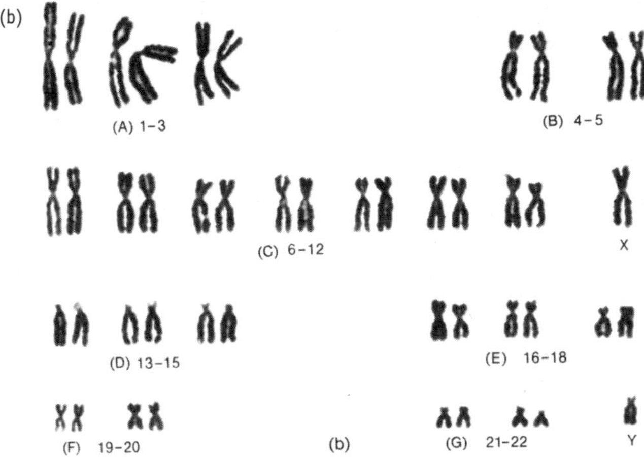

Fig. 5.3. *To create a karyotype, dividing cells are spread out on a slide so their chromosomes can be seen easily and photographed (a). Then homologous chromosomes in the photograph are paired up and the pictures are laid out neatly so the whole set can be examined (b).*

number of chromosomes that its parents have, and the number would have been doubling at each generation for millions of years. This obviously doesn't happen. Since the same numbers of chromosomes must be maintained down the generations, there is a different division process, called *meiosis*, that halves the chromosome number in gametes; that number is restored again at each fertilization. Thus, sexual reproduction must be understood as a larger cycle of events, a *sexual cycle*:

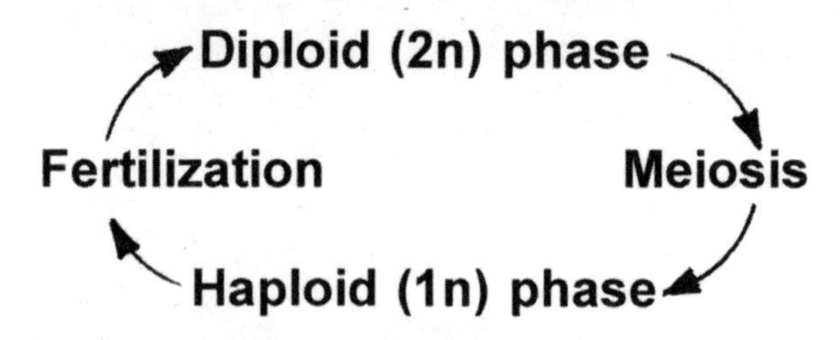

A cell that has only one set of chromosomes is *haploid* (*haploos* = simple), and one with two sets is *diploid* (*diplos* = double). The human karyotype shows two sets of twenty-three chromosomes each, so we humans are diploids. In adult gonads – testes and ovaries – some cells divide by meiosis to produce sperm and eggs, the haploid phase of the sexual cycle. These gametes each carry twenty-three chromosomes. Fertilization produces a diploid zygote with forty-six chromosomes again, which grows into another adult, and the cycle repeats.

We humans tend to think it is natural for the diploid phase to be the dominant part of an organism's sexual cycle. Many organisms, however, have only brief diploid phases, and most of their cells are haploid. In mosses, for instance, the leafy green plants are haploid, and they are much more prominent than the diploid phases, which are thin, usually brown stalks that grow out of the leafy structure. Neither phase is more "alive" than the other. We emphasize this to shed some sanity on the often raised question of just when human life begins. Human life began millions of years ago, as primitive humans separated from other primates, and it has been continuing in cyclical fashion all this time. Sperm and eggs are no less alive than embryos, and it is interesting that we worry so over the loss of diploid individuals but care nothing for all the haploid cells that are wasted every day.

The division mechanism of meiosis – centrioles, spindle, and so on – is the same as in mitosis; only the chromosomes behave differently (Figure 5.4). A cell entering meiosis has four copies of each chromosome because it is diploid – it already has a pair of each chromosome type – and because each one becomes doubled during the S phase that just precedes meiosis. After two cell divisions, it will produce four haploid cells, each bearing one copy of each chromosome. Let's take the process step by step.

As the cell begins the first meiotic division (meiosis I), homologous chromosomes somehow attract each other and pair with each other. As chromosomes become visible in prophase I, one can often distinguish irregular bumps and constrictions along the

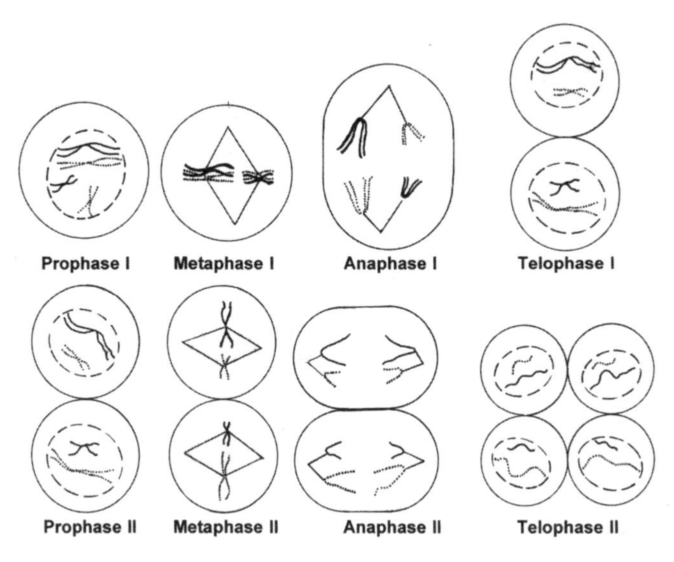

Fig. 5.4. *The general process of meiosis in a cell with only four chromosomes; one of each pair of homologs is drawn as a solid line, the other as a dotted line. (a) Prophase I: chromosomes become visible and homologs begin to pair with one another. (b) Metaphase I: the paired chromosomes line up in the middle of the cell. (c) Anaphase I: from each pair of chromosomes, one complete homolog moves toward each pole; notice that centromeres are still undivided, so the chromatids of each chromosome remain joined. (d) Telophase I: initial division is complete. (e) Prophase II: chromosomes become visible again, just as they would in a mitotic division. (f) Metaphase II: chromosomes again line up in the middle of the cell. (g) Anaphase II: now the chromatids separate from each other and move toward opposite poles. (h) Telophase II: division is completed, and there are now four haploid cells.*

strands, and as the chromatids of homologs twist around each other, these features tend to line up exactly. The chromosomes become even more compact and fatter, and now they start to pull away from each other; eventually they are held together only at points called *chiasmata* (singular, *chiasma*) where their chromatids seem to be tightly twisted around each other or even fused together.

At metaphase I, the paired chromosomes lie in the middle of the cell. They start to move toward opposite poles at anaphase I, but here each chromosome separates from its homolog while its two chromatids remain together. (This is a critical difference from mitosis, in which chromatids separate from each other during anaphase.) The chromosomes reach their poles at telophase I to be enclosed in new nuclear envelopes, and the cell divides in two. Each daughter cell – which is now haploid – goes through a brief interphase (*without* more DNA synthesis) before beginning a second division.

Meiosis II is essentially a mitotic division. The chromosomes, which still have two chromatids each at prophase II, gather along the equatorial plate at metaphase II. At anaphase II, the chromatids finally separate from each other and migrate to opposite poles, and new nuclei form around them at telophase II. There are now four haploid products, each with one copy of each chromosome. (In humans, each sperm or egg has twenty-three chromosomes. The X and Y chromosomes behave like homologs, so each sperm has twenty-two autosomes plus either an X or a Y. Try drawing them out to see for yourself.)

In animals, meiosis is slightly different in males and females (Figure 5.5). In *spermatogenesis*, the production of sperm, a single cell called a *primary spermatocyte* divides into two *secondary spermatocytes*, and these produce four *spermatids*; each spermatid matures into into a spermatozoon, with its peculiar head and long flagellar tail, in a process called spermiogenesis. Eggs, or *ova* (singular, *ovum*) are produced through a comparable meiosis in *oogenesis*, but whereas meiosis in males produces four sperm from each primary spermatocyte, the cytoplasmic divisions are very different in oogenesis, producing only a single egg. A human egg weighs only one twenty-millionth of an ounce, but its diameter is seventy times that of the sperm head. The function of a sperm is simply to carry a nucleus into an ovum. The ovum is large because it is going to develop into an embryo with no increase in mass during

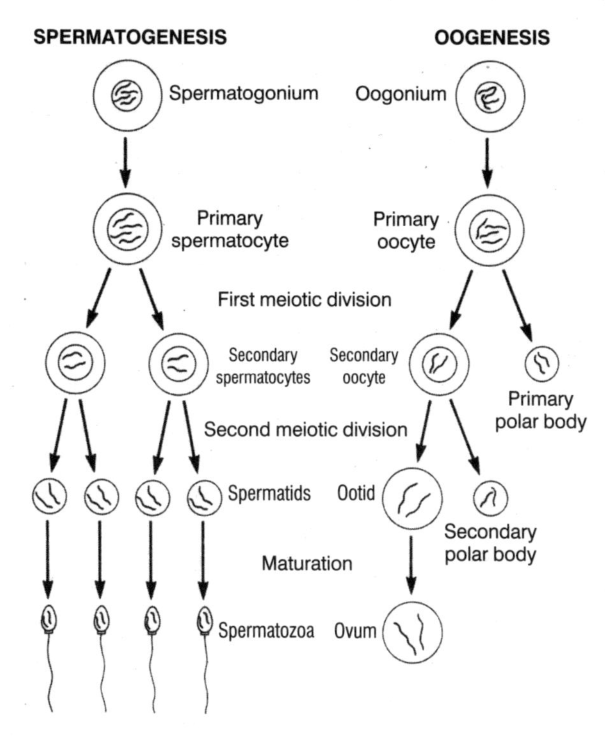

Fig. 5.5. *Spermatogenesis produces four viable products from each initial cell, but oogenesis only one. At the division of a primary oocyte, one set of chromosomes is removed in a minute polar body, which may divide again but has no further function. At the second division, another polar body is formed. In this way, the large egg is made haploid with the loss of only a little of its cytoplasm.*

its first divisions; furthermore, the ovum is a reservoir of yolk-like material that provides nourishment for the developing embryo. In a *primary oocyte*, or prospective egg, meiosis occurs near the edge of the cell, and at the end of the first meiotic division one daughter nucleus is pinched out as a tiny bud called a *polar body.* Both nuclei divide again in meiosis II, and again one of the two nuclei in the oocyte is pinched out. Thus, three polar nuclei and one egg nucleus are formed, so oogenesis produces only one egg from each oocyte.

A mature human male produces millions of spermatozoa daily. In contrast, a female releases one egg approximately every twenty-eight days for about forty years. So a woman will produce only 400–500 eggs in her lifetime. All the cells capable of forming eggs – about two million – are already present in a baby girl when she is born. This number decreases to about 300,000 by puberty, so still only a small fraction of the oocytes will ever develop into mature ova and be released.

meiosis explains mendel

It is now a commonplace of biology that genes reside on chromosomes, though we shall show in the next section just how this identification came to be made. Looking more closely at meiosis, we can see how it accounts for the basic Mendelian principles of inheritance. Mendel's first principle, the law of segregation, says that every individual carries two factors for a given trait but that the gametes carry only one of these factors. This clearly follows from meiosis. Every person – or pea plant – is a diploid with homologous chromosomes that carry the same genes, or perhaps different alleles of the same genes. To return to our former example of tasting or non-tasting the chemical PTC, somewhere among our twenty-three pairs of chromosomes is one that carries either allele *T* or allele *t*. In a heterozygote *Tt*, these homologs pair during early meiosis, but they separate at the first anaphase so each resulting gamete carries either *T* or *t*.

Mendel's second principle, the law of independent assortment, governs the inheritance of two factors simultaneously. In fact, genes will only assort independently if they are on different chromosomes; we will deal with the inheritance of linked genes – those on the same chromosome – in Chapter 8. The critical point is that each chromosome pair separates independently of the others. Suppose an organism has only two pairs of chromosomes, bearing alleles *A* or *a* on one, alleles *B* or *b* on the other. In a double heterozygote, *Aa Bb*, at metaphase I, all the chromosome pairs line up in the middle of the cell:

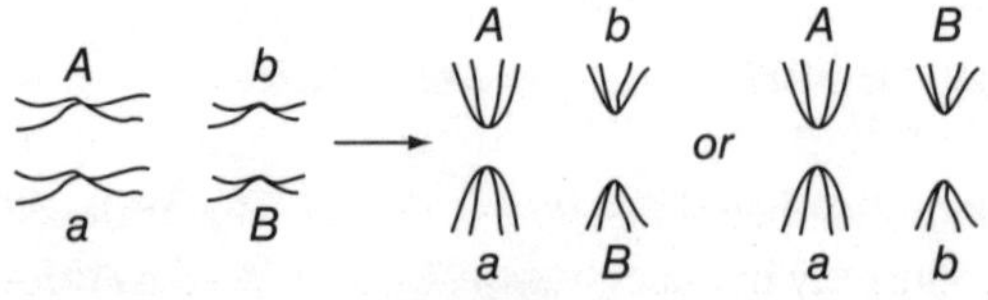

At anaphase I, sometimes alleles *A* and *B* will move together into one daughter cell, alleles *a* and *b* into the other. But it is equally likely that the chromosomes will separate with *A* and *b* in one cell, *a* and *B* in the other. Thus, many meiotic events will create equal numbers of the four types of gametes: *AB*, *Ab*, *aB*, and *ab*. This accounts for the patterns of inheritance we examined in Chapter 4.

the location of genes

The basic events of both meiosis and mitosis were well established by the late nineteenth century. Although we can now see that they are elaborate mechanisms for distributing chromosomes properly into daughter cells, no one understood their biological role until the twentieth century. With the rediscovery of Mendel's principles in 1900, Theodor Boveri and Walter Sutton independently recognized that the behavior of genes, according to Mendel's two laws, closely paralleled the movement of chromosomes in meiosis. In 1902, they proposed the Sutton–Boveri chromosome theory of heredity, that the hereditary factors defined by Mendel are on chromosomes. They pointed to these parallels:

Genes	Chromosomes
1. Occur in pairs, one of each pair coming from each parent.	1. Occur in homologous pairs in diploids resulting from the union of haploid gametes from each parent.
2. Alleles of a pair segregate from each other into different gametes.	2. Pair and separate into different gametes in meiosis.
3. Different pairs assort independently.	3. Different pairs appear to distribute themselves independently.

The remarkable parallel in the behavior of chromosomes and genes is not proof that genes are on chromosomes, and many biologists, including geneticists, were highly skeptical of Sutton and Boveri's proposal. Various experiments and observations during the early years of the twentieth century soon established Sutton and Boveri's hypothesis.

sex chromosomes

Even in antiquity people observed that certain traits occur most often or exclusively in males but were transmitted by their mothers.

Hemophilia, an inability to form blood clots properly, is the best-known example. To hemophiliacs, or "bleeders," a minor cut or bruise threatens uncontrollable bleeding. The ancient Hebrews demonstrated their observation that the deficiency is hereditary by decreeing that circumcision, an otherwise essential ceremony in their society, be avoided for infant sons when two earlier brothers had died of bleeding after the operation. Furthermore, by the twelfth century rabbinical scholars had recognized that hemophilia, which affected males almost exclusively, was paradoxically transmitted by females, not males. Charles Darwin also noticed this peculiar type of inheritance. In 1875, he wrote about a Hindu family in which ten men, over four generations, had only a few small teeth, very little body hair, early baldness, and excessively dry skin. No woman in the family was ever affected, but they could transmit the syndrome to their sons, whereas a man never transmitted it to his sons.

The explanation for such patterns of heredity rests with the gender difference in chromosomes that is apparent in karyotypes: females having two X chromosomes while males have one X and a much smaller Y chromosome, which acts as its pairing partner during meiosis. All the eggs that females produce contain a single X, whereas males produce two types of sperm in equal numbers, half carrying an X and half carrying a Y. This accounts for the equal sex ratio, since every egg is fertilized by an X-bearing sperm, producing a female XX zygote, or by a Y-bearing sperm, producing a male XY zygote. It is ironic that a child's sex is determined by the sperm, not the egg of the mother, for women are still commonly blamed for failing to give birth to a child of the desired sex. Many monarchs have divorced their wives for failing to produce a male heir. However, in other animals – including amphibians, birds, butterflies, and moths – it is the egg that determines sex; males have two identical sex chromosomes called Z and females carry two different ones, a W and a Z.

The human Y chromosome carries few known genes. A small region of the chromosome called SRY determines maleness by converting the gonads into testes instead of ovaries; any zygote with a Y chromosome develops into a male while a zygote without one becomes a female. Any trait determined by a gene on the Y chromosome would show a distinctive pattern of inheritance, passing only from father to son. About the only well-documented case involves a phenotype referred to as "hairy ear rims." The trait often appears late in life and is variable in expression, so the inheritance patterns are not straightforward, although the trait is evidently inherited as expected from father to son.

However, many traits are X linked – their genes are carried on the X chromosome – and their location is easy to identify because of their distinctive pattern of inheritance. One well-known human trait, red–green color-blindness, makes a good example; let's designate an X chromosome with the mutant allele by X^c and a chromosome with the normal allele by X^+. Because the color-blindness mutation is recessive, a woman who is heterozygous for the gene ($X^+ X^c$) has normal vision. However, a man who receives the mutant allele on his sole X chromosome ($X^c Y$) has no normal allele to cover up the mutation and is therefore color-blind. A color-blind man gives his X^c chromosome to all of his daughters, making them (usually) heterozygous carriers of the trait. (None of his sons receives the trait from him, since they inherit his Y chromosome.) A heterozygous woman gives half her sons the normal X^+ chromosome and half the X^c chromosome. Color-blind women are rare, since they must have a color-blind father and a heterozygous mother, and they have only a chance of ½ of receiving their mother's X^c chromosome.

X-linked traits can therefore be identified in a pedigree because they pass from women to half their sons and from men through their daughters, skipping a generation, to their grandsons. This pattern of inheritance is characteristic of several hundred human characteristics, including some forms of baldness and Duchenne's muscular dystrophy. One of the best-known cases is the inheritance of hemophilia in the royal lines of Europe (see Chapter 14).

nondisjunction of chromosomes

Normal males and females have well-known phenotypes determined by their XY or XX chromosome sets. Occasionally people appear with unusual numbers of sex chromosomes who have an abnormal development of their gonads called *gonadal dysgenesis* and other unusual sexual characteristics. Two syndromes bear the names of the physicians who first identified them. *Klinefelter syndrome* is seen in boys who are typically tall, with gynecomastia (breast development), below-average mentality, and small testes. In 1959, Jacobs and Strong recognized that Klinefelter syndrome was correlated with the XXY condition – an extra X chromosome.

A second type of gonadal dysgenesis called *Turner syndrome* occurs in females. Turner females have no ovaries, are short in stature, have underdeveloped secondary features (such as their breasts), and a webbing of flesh in the neck region. Turner females are XO – they have only a single X chromosome, the "O" indicating the lack of a chromosome. Since these females are hemizygous for the X chromosome, they can exhibit recessive phenotypes, such as color-blindness, which are usually expressed in males. About one in every seven hundred births is XXY, and about one in every 2500 is XO. In addition, XXX females (called *triple-X*) occur about once in every thousand births and appear normal, although they may have reduced mental capacity.

How do the XXY and XO conditions arise? The specific cause is still not known, but it is clear that on rare occasions during meiosis paired chromosomes do not separate, or *disjoin*, properly (Figure 5.6). This

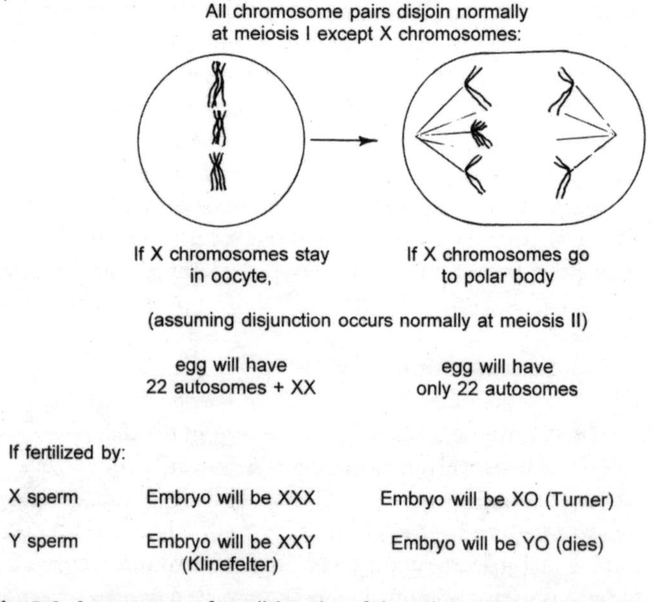

Fig. 5.6. *Consequences of nondisjunction of the X chromosomes in an oocyte at meiosis I, followed by fertilization with an X-bearing or Y-bearing sperm. Nondisjunction at meiosis II (not shown) could add even more X chromosomes.*

event is called *nondisjunction*. It can occur in either sex and at either meiosis I or II, or both. Nondisjunction can produce gametes that have two sex chromosomes (XX, YY, or XY), or none (called nullo), or in very rare cases even more copies. (Nondisjunction of autosomes can also lead to severe disorders, as discussed in Chapter 14.) The combination of an XY sperm with an X egg then produces an XXY offspring; a nullo sperm combined with an X egg produces an XO offspring. Fertilization of an X-bearing egg by an XX sperm produces a triple-X female, and fertilization by a YY sperm gives an XYY.

These abnormalities in sex chromosomes raise the question of how XX women and XY men can both be normal when they have different numbers of X chromosomes. There must be a mechanism to compensate for this difference and maintain a genetic balance. In 1961, Mary Lyon and Liane Russell independently presented a model for compensation of X-linked genes. They pointed out that heterozygous females often have a variegated phenotype; calico cats, for instance, are heterozygotes that have randomly distributed patches of black and yellow fur. Lyon and Russell proposed that in every cell of a developing female embryo, one X chromosome is *inactivated* at random, and the same chromosome remains inactive in all the cells derived from that embryonic cell. In a heterozygous cat, the X chromosome with the black-fur allele is turned off in some skin cells, which then develop yellow fur patches; in other skin cells, the chromosome with the yellow-fur allele is turned off, and they produce black patches. Although this pattern of X inactivation is most obvious in the fur of female cats and mice, every female mammal is a mosaic of two cell types, and any allelic difference between her X chromosomes may show up as a phenotypic difference.

X chromosomes are turned off by being condensed into compact bundles. These are called sex chromatin or *Barr bodies*, after Murray Barr, who discovered them; they can be seen in the cells of normal females. One X is left functional and the rest are condensed, so a person typically has one fewer Barr body than the number of X chromosomes. So a normal woman has one Barr body in each cell, a woman with Turner's syndrome has none, and women with extra X chromosomes have two, three, or even four. Men normally have no Barr bodies, but those with Klinefelter's syndrome show one, two, or more, depending on the number of X chromosomes in their cells.

xyy males: a genetic dilemma

In 1956, Patricia Jacobs and her associates published a report, the repercussions of which were debated for a long time. In a study of the karyotypes of inmates in the wing for mentally retarded men in Carstairs Maximum Security Hospital in Scotland, they found that seven out of 196 men (3.6%) were XYY. One phenotypic trait associated with the XYY condition is significantly greater height than normal. In a later study of the chromosomes in 3500 consecutively born male babies in a hospital, they found five XYYs (0.14%). Thus, if XYY males represent only 0.14% of the population, the 3.6% among inmates of a high-security prison suggests that such males have a tendency to commit crimes of violence. The immediate conclusion, if this is true, is that the Y chromosome carries a gene (or genes) for aggressiveness, which would cause that much more violent behavior when present in two doses. This resulted in lurid headlines featuring the Y as a "criminal chromosome."

Indeed, newspapers reported that Chicago mass-murderer Richard Speck, who bore the tattoo "Born to raise hell," was XYY. He was not. Nevertheless, several accused murderers tried to escape conviction for murder because they had an XYY genotype. By 1974, there were at least six criminal trials in which the defendant used his XYY chromosome condition as the basis for a plea of insanity: Daniel Hugon in Paris; Lawrence Hannell in Melbourne; Ernst-Dieter Beck in West Germany; Sean Farley in New York; Raymond Tanner in Los Angeles; and the case of *Millard v. State* in Maryland.

Despite the lack of conclusive evidence, highly tendentious, even inflammatory statements were made by a number of people, including geneticists. For example, Kennedy McWhirter, a lawyer and geneticist, stated that "the probability factor makes the criminal XYY a predictably dangerous person."[2] A law journal suggested

> The XYY individual is a perpetual threat since at any time he may face a situation in which he will be unable to control his behavior. Although there may be many thousands of walking "powder kegs," XYYs are among the few who are easily detectable.[3]

It was reported that "one of the country's leading geneticists" said

> We can't be sure XYY actually makes someone a criminal, but I wouldn't invite an XYY home to dinner.[4]

Even a scientist as eminent as H. Bentley Glass, past president of the American Association for the Advancement of Science, characterized XYYs as "sex deviants" to be eliminated from the population by modern techniques of fetal diagnosis and abortion.[5]

The truth of the matter is that no definite connection between the XYY constitution and criminality has been found by serious researchers. In a careful evaluation of several studies, Ernest B. Hook[6] found that the rate of XYY births in the population is about 0.1%, but XYY males appear at a rate of about 2% in "mental-penal" settings (prisons as well as facilities for retarded, disturbed, psychotic, alcoholic, or epileptic individuals). But careful study of XYY individuals show that these people have no significant tendency toward aggressiveness or criminality. Their most consistent phenotypic trait is tallness. They also tend to have severe acne and may exhibit some mild mental retardation and perhaps difficulties in socialization, such as an inclination to be impulsive. Interestingly, XXY (Klinefelter) males, who are also above average height, also appear at a greater rate in mental-penal settings – about five times that of the whole population. Thus an explanation for the prevalence of XYY and XXY males in institutions is that tall youths generally experience somewhat greater difficulties in social adjustment and authorities are more likely to view them as threatening and thus to shunt them into pathways likely to lead to penal or mental settings.

The history of the XYY case ought to serve as a warning. Scientists are usually conservative in accepting new theories, yet in this case infatuation with a new idea, bolstered by very shaky data, prompted an uncritical endorsement of the connection between criminal behavior and the XYY condition. Reasoning of this kind could do incalculable harm. To follow that logic, it has been suggested facetiously, would be to say that, since males far outnumber females in prisons, the XY genotype itself is crime prone.

To the geneticist, the real significance of conditions such as XYY is that they arise from chromosomal nondisjunction and demonstrate the correlation between particular chromosome constitutions and phenotypes. Genes are indeed located on chromosomes, which is why gene expression and chromosomal conditions go together.

THE LONG CONCERN OVER SEX DETERMINATION

Early attempts to understand reproduction were strongly motivated by the desire to predict, and even to ensure, the sex of an infant. Throughout the ages many peoples have preferred to give birth to boys. Since many societies believed semen to be the sole transmitter of heritable characteristics, a son was the only means to leave a biological legacy – a kind of immortality, often symbolized by the continuation of the family name. In most societies, property went only to males, so many parents wished for sons to keep their accumulated wealth from being distributed to more remote male kin. In societies where chieftainship, kingship, the shamanic role, priesthood, or other social positions were awarded through succession, the entire community prayed for sons to avoid the gaps that lack of heirs would create in their social organization. Some families coveted sons in anticipation of the dowries that new daughters-in-law would bring. Others longed for sturdy hunters and warriors to ensure the survival of their clan. Daughters, on the other hand, were not only extra mouths to feed but, in a number of societies, also required large dowries, which could ruin the man who was unfortunate enough to have several daughters. For all these and other reasons, the birth of a son in early societies became a matter of intense pride, a custom that persists today in most societies.

From time to time we come across a society that treasures its daughters, although they sometimes become proud *possessions* to be handed from male to heir along with other material property. In some societies, females do inherit property, and in these societies parents rejoice at the birth of girls. Whatever the preference of a society, the wish to ensure that a child be of the desired sex led to numerous fascinating "rules," which were attempts at a primitive genetic engineering.

The ancient Egyptians had no rules on how to conceive one sex or the other, but they had methods to discover what sex could be expected. They taught that a woman's urine poured daily onto separate bags of barley and wheat would cause the barley to germinate if she were carrying a girl and the wheat if it were a boy. No germination indicated the woman was not pregnant. This seemingly bizarre method may have had some real basis, since the urine of pregnant women does contain substances that stimulate the growth of certain plants.

The early Hebrews were anxious to produce boys, since women had little status in their society and could not inherit property. The Hebrews devised theories about ensuring a son, such as the belief that a man who burned his neighbor's grain would leave no son to be his heir. But if he placed his marriage bed in a north–south position, says the Talmud, he would be likely to father sons. The Hebrews shared with Arabian and Hindu cultures the belief that the birth of a son is a joyous occasion, so a woman who has a happy, untroubled pregnancy was presumed to be carrying a male baby.

A pregnant woman's nutrition could also determine the sex of the child. In ancient India, the *Atharva-veda* (written before 700 BC) urges a mother who wishes a son to drink *garbhakarana* – a drink made of rice and sesame seeds boiled in water – on the fourth day of menstruation. The ancient writings also contain detailed directions for preparing food to produce particular characteristics of the son, such as the color of his complexion and the number of Vedas he would become learned in.

During the time of Homer, the Greeks coveted daughters. The sons, in fact, were often exposed to the elements and allowed to die, because daughters brought cattle as the price of marriage. Several centuries later, however, the preference had swung toward sons, the only offspring who could inherit by traditional law. Daughters could receive the dowry only when they married. If a man died without heirs, the courts awarded his wife and all his property to the next of kin. Naturally, people in that society were anxious to predict the sex of their child and sought to discover methods of sex determination. Hippocrates taught that a woman who had a healthy coloring during pregnancy would bear a son. He also believed that a fetus vigorous enough to move about at three months gestation must be a son; one who took four months to move must be a girl. And if the mother develops freckles, her child will be a daughter. (In most folklore, however, freckles indicate a son.)

Hippo of Rhegium thought sex was determined by the male, which it is, but he felt that if the semen was thick it would produce a boy and if thin a girl. To produce a son, the Greeks

(continued overleaf)

(continued)

therefore made love when the north wind was blowing, because the north wind, as everyone knows, thickens and activates the semen. For a daughter, a south wind produced the necessary watery, feeble consistency. Empedocles, on the contrary, thought a warm uterus was the factor that produced boys.

Many theories of sex determination come from the ancient belief that the right side of the body is more important than the left, perhaps because of the prevalence of right-handedness. Since it was widely believed that men have two testicles in order to produce sperm for the two sexes, the right testicle must then be the one to form sperm that could produce boys. Many men believed that tying off their left testicle would produce a boy. Aristotle argued against this idea, but it proved lasting: many a French nobleman of the eighteenth century actually had his left testicle removed in the hope of producing an heir.

The mother's dreams have often been considered significant. In India, dreams of men's food are believed to predict a son; in Slavic countries, knives and clubs signify the same, while dreams of spring or parties prophesy a girl. In Germany's Spessart mountains, some husbands still take an ax to bed with them for a little help in conceiving a son and leave it in the woodshed when they prefer a girl. In tenth-century Japan, pregnant women hunted and fought to encourage the fetus to become a boy. And until recently, Japanese entertained the remarkable belief that when a man calls to his wife while she is on the toilet, if she turns suddenly to the left, the baby will be a girl. An old Chinese rule instructs that if the mother feels the baby's hand move to the left after the seventh month, she will bear a boy; and if after the eighth month it moves to the right, it is a girl.

Today people still believe all kinds of folklore for predicting sex, including the rate of fetal heartbeat, frequency of maternal vomiting, freckling pattern in the mother, and the quality of her milk. Now it is possible to make an unambiguous determination of fetal sex, and techniques are being developed to control the sex at the time of conception. In the past, the desirability of controlling a baby's sex was taken for granted. Today, when the possibility lies within our grasp, we have the sudden responsibility of facing the pros and cons involved in fulfilling this age-old wish.

chapter six

the function of genes

Though our course of study has given us considerable understanding of how chromosomes and genes are transmitted from generation to generation, we have not yet gained a clear insight into what genes actually do. We're now ready to delve into that question.

genes and metabolic disease

Although humans are poor subjects for the study of heredity because good breeding data are so difficult to obtain, the first clues to what genes actually do came from observations on humans. In the early twentieth century, an English physician, Archibald Garrod,[1] was intrigued by certain disorders in people for which there was a clear hereditary history, and he began to compare these people biochemically with normal people. One of these disorders was an autosomal recessive condition, *alcaptonuria*, whose most striking effect is that the urine of affected people turns dark (red through black) upon exposure to air. Parents often notice this in the diapers of affected babies. Affected people excrete a great deal of homogentisic acid (also called alcapton), which turns dark when it reacts with oxygen. The cells of normal people convert homogentisic acid into maleylacetoacetic acid, so alcaptonuria results from the absence of the enzyme that normally catalyzes this reaction. Remember that the components of an organism are made or broken down through metabolic pathways, which are series of chemical reactions catalyzed by specific enzymes. Garrod therefore called alcaptonuria and some similar disorders "inborn errors of metabolism".

The nature of these disorders suggests that a gene carries the information for an enzyme. Normal people, carrying the allele *Al*, can make the key enzyme that converts homogentisic acid to maleylacetoacetic acid. Alcaptonurics are homozygous for the recessive allele *al*, which carries defective information. Heterozygotes have one copy of the normal allele and therefore have normal metabolism. Similarly, Garrod pointed out that an albino lacks melanin, the pigment present in skin, hair, and iris cells of normal people. Melanin is synthesized enzymatically from tyrosine by a series of reactions, and albinos lack an enzyme somewhere along the pathway. This is the critical connection between genes and other components of an organism.

We now know several hereditary defects in humans that result from blocks at specific points in the pathway leading to melanin and to maleylacetoacetic acid (Figure 6.1). So we might conclude from observations like Garrod's that genes control enzyme production, and thus metabolism in general. Certain metabolic defects have shaped the flow of history. For example, porphyria results from an inability to metabolize compounds called porphyrins, leading to insanity. The erratic behavior of King George III of England – probably a factor in precipitating the American Revolution – was the result of porphyria inherited from Mary, Queen of Scots.

genes and enzymes

In 1944, George Beadle and Edward Tatum verified Garrod's speculations in a striking way by using the bread mold *Neurospora*. (If you look at a moldy piece of bread and notice a bright orange growth, that's *Neurospora.*) *Neurospora* can be grown on a simple *nutrient medium*, a mixture of sugar, the vitamin biotin, and a few salts in water. From these substances, the mold synthesizes all of its complex components, and an organism that can grow in this way, without any nutritional supplements, is called a *prototroph*. Beadle and Tatum irradiated *Neurospora* spores to induce mutations and then screened them to find the kinds of mutants they could use,

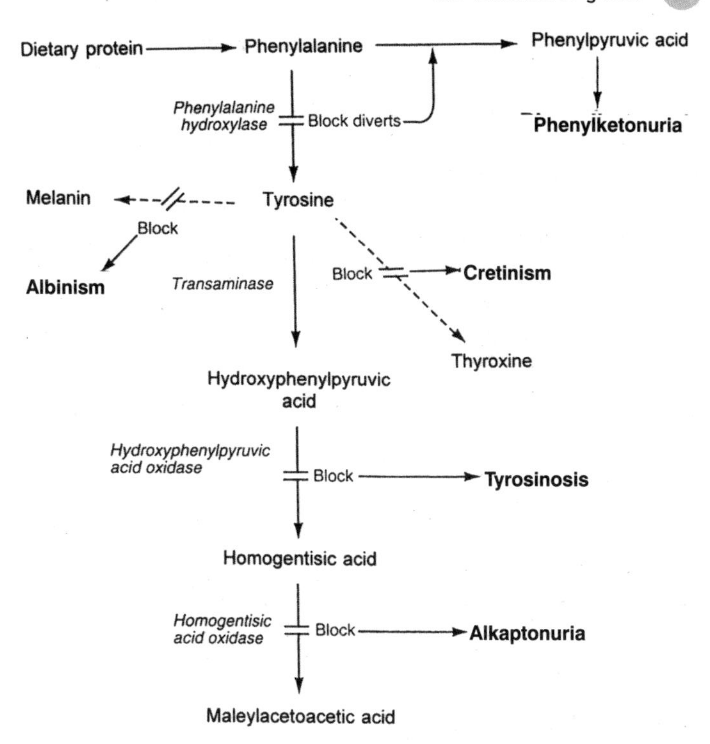

Fig. 6.1. *Metabolic pathways of tyrosine and phenylalanine metabolism in humans. Absence of enzymes at different points results in specific hereditary defects.*

which are called *auxotrophs*; an auxotrophic mutant is unable to produce at least one of its own components, so it can only grow if that material is added to the growth medium.

Beadle and Tatum focused on mutants that specifically required the amino acid arginine to be added to the medium. Arginine is an essential component of all proteins, and normal *Neurospora* can synthesize it from sugars and other substances in the medium. But the mutants have defects in the metabolic apparatus for synthesizing this one compound, so without this additive they could not grow. Arginine is similar in structure to two other amino acids, ornithine and citrulline, so Beadle and Tatum surmised that some mutants might grow by using these other compounds instead of arginine; in this way, they divided their arginine auxotrophs into three distinct classes. Class I mutants would grow only on a medium supplemented with arginine; class II mutants could grow when provided either with arginine or citrulline; class III mutants could grow with arginine or citrulline or ornithine. From these results, Beadle and Tatum reasoned that (1) a gene carries the information for an enzyme, (2) each mutant is defective in one gene and therefore in one enzyme, and (3) the biosynthetic pathway leading to arginine production has the structure:

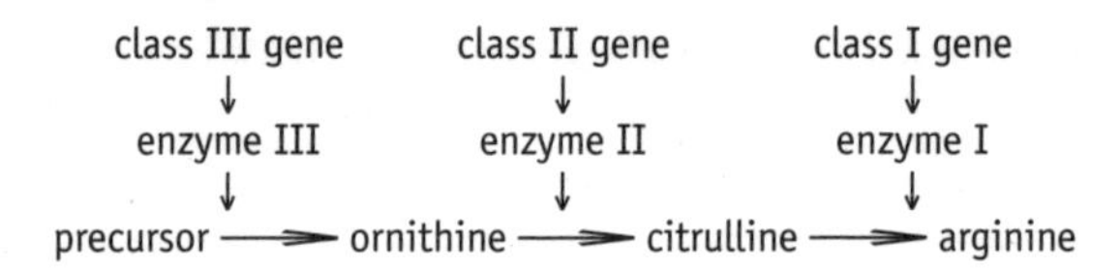

Each reaction is controlled by a specific enzyme, and each mutant is apparently missing just one enzyme and is thus blocked at a specific step. No mutant can grow on a compound that comes *before* its block; class II mutants cannot grow on ornithine, for instance. But if a mutant strain is provided with a compound that comes *after* the blocked step, it has the other enzymes necessary to convert this compound to arginine, so it can grow. Thus, class III mutants cannot convert the precursor to ornithine, but when provided with ornithine they can convert it to citrulline and then to arginine. Later biochemical studies showed that these reactions do indeed proceed in the sequence inferred from studies of these mutants.

Based on these results, Beadle and Tatum devised the "one gene, one enzyme" hypothesis: that the function of each gene is to carry the genetic information for a single enzyme. In fact, they had hit upon the crucial notion that genes carry the information to build proteins; enzymes are one very important type of protein, but, as we showed in Chapter 3, there are many other important types of protein. A more general, modern version of the Beadle and Tatum principle is that each gene codes for one polypeptide chain. A polypeptide is simply a chain of amino acids. Some proteins are made of a single polypeptide, others of two or more different polypeptides.

proteins and information

Since genes control the structure and synthesis of proteins, we need to take another look at protein structure. As we noted in Chapter 3, proteins are the most versatile molecules in organisms. They are essential components of all membranes; they make the bulk of the ribosomes, which are factories for making more proteins; they form many of the structures that move and shape cells, and are the active components of the muscle fibers that pull on one another so we can move our bodies. They make large, obvious structures such as hair, the skin surface, tendons, and (in combination with crystals of calcium magnesium phosphate) bones. Some proteins are hormones, such as insulin, which carry chemical messages in the body from cells of one type to cells of another type. And, most important, they are the enzymes that catalyze chemical reactions.

Remember that proteins are linear polymers of twenty kinds of amino acids, which can form an enormous variety of sequences. Two amino acids are joined by a *peptide linkage* in a chemical reaction that releases water as a by-product, and a polypeptide is made of many amino acids polymerized in this way. Machines can now quickly determine the sequence of amino acids in a protein. Sequence analysis has shown that each kind of protein has a unique sequence, called its *primary structure*; all protein molecules of one specific kind have the same specific sequence. For example, to use the three-letter abbreviations for the amino acids, the polypeptide Glu–Gly–Pro–Trp–Leu–Glu–Ala–Tyr–Gly–Trp–Met–Asp–Phe is the hormone gastrin, which helps to regulate digestion, and the polypeptide Tyr–Gly–Gly–Phe–Met is an enkephalin, a substance that acts like an opiate (painkiller) in the brain. (We always write these sequences from the *amino terminal* (N-terminal) end, which has a free amino group, to the *carboxyl terminal* (C-terminal) end, which has a free acid group). Most proteins, however, are much larger molecules made of a few hundred amino acids, and altering just one of them can make a molecule with a very different structure, often one that has lost its function.

The primary structure basically determines the properties of every protein, including the way the long chain of amino acids folds up into a specific shape (Figure 6.2). Attractions between the atoms

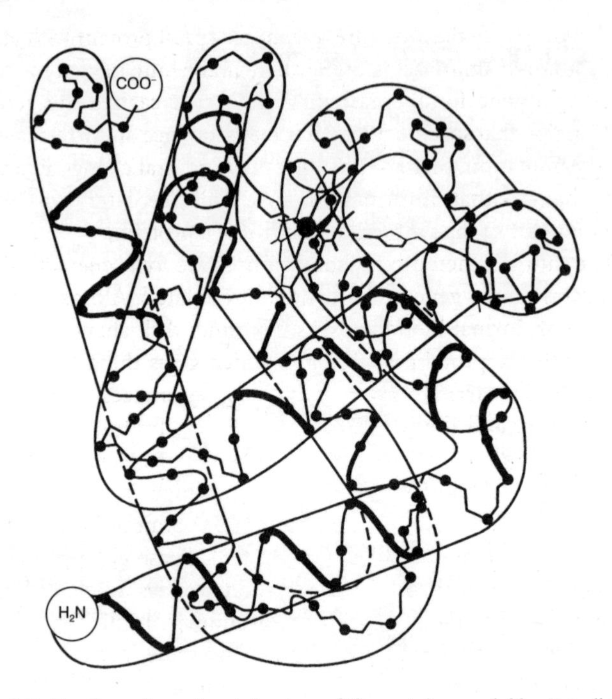

Fig. 6.2. *The three-dimensional structure of the protein myoglobin. Not all the atoms of the molecule are shown; the heavy line traces the backbone of the peptide chain, which forms several short helical segments joined by less regular regions.*

in each protein molecule make the polypeptide chain fold up into a rather rigid structure with a form appropriate to its particular function – for instance, with a small pocket that is the active site where an enzyme catalyzes a chemical reaction. Also, many functional proteins consist of a tight cluster of two or more polypeptides.

In talking about genes, we used the term "information," especially "genetic information." Remember (Chapter 3) that we require information to specify one out of a range of possibilities, and this has particular genetic meaning in relation to protein structure. In writing the sequence of the hormone gastrin above, we specified one sequence out of 20^{13} possible sequences – about one out of 8×10^{16}. That requires a lot of information, though nowhere near the amount needed to specify the structure of a more typical protein with about three hundred amino acids. Somewhere in the human genome there must be a gene for the gastrin structure; it carries the genetic information that specifies that particular sequence of thirteen amino acids. A mutation, then, is most likely a structural change in a gene that changes that information. (We will see later that some mutations affect the synthesis or control of a protein.)

The human hereditary condition *sickle cell anemia* nicely illustrates how a gene and a protein are related. Almost half the volume of normal blood consists of round, disk-shaped red cells filled with the protein *hemoglobin*, which gives them their color. Hemoglobin carries oxygen, which all cells need for normal metabolism, and the red blood cells slip easily through the fine capillaries in all tissues, where they release their oxygen. Most of this hemoglobin is hemoglobin A (Hb A), which is made of four polypeptides: two called alpha (chains of 141 amino acids) and two called beta (146 amino acids) (Figure 6.3). These polypeptides are the globin part of the molecule; each one carries a heme group, a large ring-shaped molecule with an iron atom in its middle. The iron is actually the part that binds an oxygen molecule, and we need iron in our diets primarily for making heme groups.

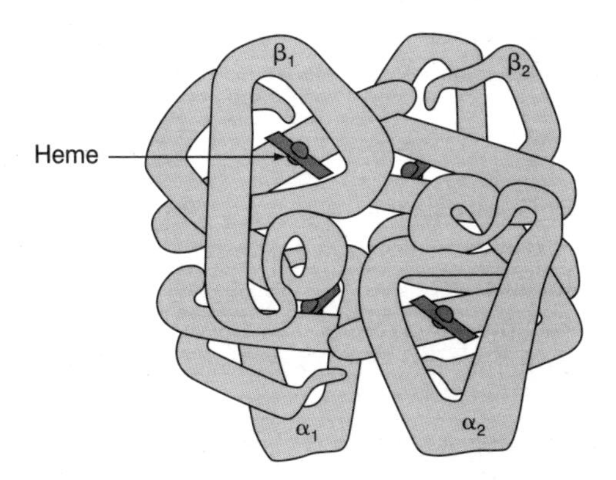

Fig. 6.3. *The three-dimensional structure of hemoglobin. Each molecule consists of four polypeptide chains, two alpha chains and two beta chains, each of which is very similar to one chain of myoglobin. The four fit together where their shapes are complementary to make a quaternary structure. Each chain contains a large, flat heme ring with an iron atom in its center; this is where oxygen molecules are bound.*

Sickle cell anemia is caused by an abnormal hemoglobin, Hb S, encoded by a mutant allele *HbS*. Red cells filled with Hb S assume bizarre shapes (Figure 6.4), some of which look like sickles. These cells don't flow easily and often jam up in small vessels to cause extreme pain, organ and tissue damage, and death. The affected person is anemic because the sickle cells last less than a third as long as normal cells. Heterozygotes, $Hb^{A}Hb^{S}$, have *sickle trait*. Their mixture of Hb A and Hb S makes them generally healthy, and less than one percent develop the anemia. The sickle cell allele is particularly common in Africa and other places where malaria is endemic (and also among African-Americans) because *HbS* produces resistance to this disease, so heterozygotes have a selective advantage.

In 1956, Vernon Ingram showed that Hb A and Hb S differ only in one amino acid in the beta chain: the sixth position from the amino terminus, which is normally glutamic acid, is changed to valine in Hb S:

Hb A: Val–His–Leu–Thr–Pro–*Glu*–Glu–Lys–
Hb S: Val–His–Leu–Thr–Pro–*Val*–Glu–Lys–

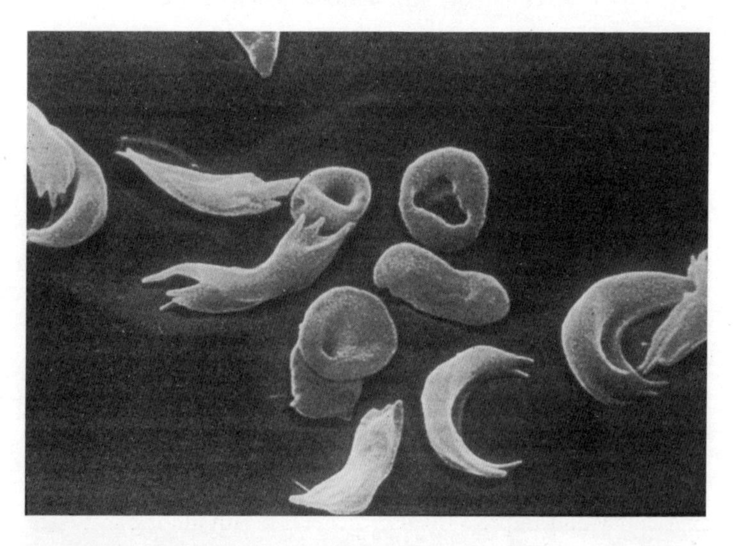

Fig. 6.4. *A scanning electron micrograph of red blood cells from a person with sickle-cell anemia shows the bizarre shapes they assume, in sharp contrast to the smooth, round shape of normal cells. Picture courtesy of Omikron/Science Photo Library.*

So out of the whole sequence of 146 amino acids in this protein chain, only one amino acid makes all the difference between normal and diseased people.

Since Ingram's initial work, the entire amino acid sequences of both hemoglobin chains have been determined, and many hereditary alterations in them have been recorded (Figure 6.5, p. 102). In each case, the allelic difference results in a protein that is identical to the normal except for one amino acid in one location, or sometimes a short stretch of deleted amino acids. This correlation – one mutation, one amino acid change – reinforces the concept that each gene dictates the exact amino acid sequence of a single protein type, and the gene must consist of a series of units – each of which can be mutated – that specify each amino acid. A mutation typically produces a single amino acid substitution, altering the protein so it has a slightly different shape and generally functions differently. Other kinds of mutations delete one or more amino acids, and some mutations change a long series of amino acids or create only a fragment of the normal protein.

modification of hereditary disease

As the science of genetics developed early in the twentieth century, the program called *eugenics*, aimed at breeding improved people, held a certain vogue, as discussed in Chapter 15. However, our knowledge of what genes do opens up possible ways to correct the phenotypes of hereditary defects – what the microbiologist Joshua Lederberg has termed *euphenics*. Some obvious euphenic procedures require no insight into the molecular nature of a defect – for example, prescription spectacles for myopia, surgical correction of cleft palate, amputation of the extra digit of a person with polydactyly, or blood transfusions for hemophiliacs. A most dramatic example of euphenics resulted from Banting and Best's 1922 discovery of insulin, which resulted in a greatly increased life expectancy and improvement of general health among diabetics based only on the knowledge that diabetes is caused by the absence of insulin. But euphenic intervention can go much farther with the precise and detailed understanding scientists now have of how genes work.

Position		Mutant
1	Val	
2	His→Tyr	Toguchi
3	Leu	
4	Thr	
5	Pro	
6	Glu→Val	S
	↘Lys	X, C
7	Glu→Lys	C Georgetown, Siriraj
	↘Gly	G
8	Lys	
9	Ser→Cys	Porto Alegre
10	Ala	
11	Val	
12	Thr	
13	Ala	
14	Leu→Arg	Sogst
15	Trp	
16	Gly→Arg	D Bushman
	↘Asp	J Baltimore, Trinidad
17	Lys	
18	Val	
19	Asn	
20	Val	
21	Asp	
22	Glu→Lys	G Saskatoon
23	Val→---	Freiburg
24	Gly	
25	Gly	
26	Glu→Lys	E
27	Ala	
28	Leu→Pro	Genova

Fig. 6.5. *The sequence of the first twenty-eight amino acids of the normal beta chain of human hemoglobin showing the positions of amino acid substitutions by different mutations. Some of the mutants are given letter names and others are named for the places where people with the variant were discovered. One mutation is a deletion that removes an amino acid.*

dietary control of phenylketonuria

In the 1930s, the mother of two mentally retarded children in Norway noticed that they had a peculiar musty odor. She took the children to a physician-biochemist named Følling who put them through a variety of tests. He found that their urine turned ferric chloride bright green, a result of a chemical reaction with phenylpyruvic acid, which is absent from the urine of normal people. This observation led to the recognition of *phenylketonuria* (PKU) as an inborn error of metabolism.

An enzyme (phenylalanine hydroxylase) normally converts the amino acid phenylalanine to tyrosine (see Figure 6.1, p. 95). Homozygotes for an autosomal recessive mutation, *p*, cannot carry out this reaction, so phenylalanine accumulates in the blood and spills over into the urine; *pp* children may have up to thirty times the normal amount of phenylalanine in their blood. The excess phenylalanine is shunted through an alternative pathway to form phenylpyruvic acid, a highly toxic compound that damages the developing brain of an infant. Consequently, *pp* children usually suffer from extreme mental retardation, one of the phenotypes of PKU.

PKU children occur at a rate of about one in every ten thousand births, and in North America from 1955 to 1964 diapers containing ferric chloride were used to test newborn children for PKU. However, the diaper spot test proved to be too crude a diagnostic tool, detecting only about 50–60 percent of all PKU children, so it is no longer used. Now PKU is detected very simply at birth by taking a small blood sample (usually from the heel) on a piece of blotting paper. The dried blood spot is then punched out and placed on a petri plate containing nutrient agar medium, the bacterium *Bacillus subtilis*, and the compound β-2-thienylalanine, which inhibits growth of the bacteria. Phenylalanine overcomes the inhibitor and allows the bacteria to grow, so excess phenylalanine in the blood sample will diffuse out and produce a halo of bacterial growth. This simple procedure allows rapid and accurate assays of many blood samples.

Children initially diagnosed as possible PKUs are then retested, and if the initial diagnosis is reconfirmed they are placed on a special diet with greatly restricted amounts of phenylalanine. Low-phenylalanine food is not a gourmet's delight, and as the first group of children on the diet grew older, they began switching to better-tasting, ordinary foods with no apparent ill effects. So as long as children are fed low-phenylalanine diets in the first few years of life, they achieve normal intelligence. However, it is of some concern that though the mental ability of PKU homozygotes varies considerably, most of the children of PKU women are retarded regardless of their genotype. Since these women generally marry normal men, their children are heterozygous and therefore should be normal. The explanation is that the uterine environment provided by the PKU mother contains too much excess phenylalanine for the small circulation of the developing baby to handle, and it permanently damages the baby's brain.

Mass screening for PKU began experimentally in Britain in the mid 1950s. As techniques for diagnosis were refined, tests on larger scales were adopted. In 1964, a concerted push to legislate compulsory PKU screening in the United States was initiated, and within two years forty-three states required such tests for all newborn infants. Now, in the U.S. alone, over three million infants are screened for PKU each year. A study of the results of mass PKU screening showed that only 183 treatable PKU children were detected at a cost of 5–10 million dollars annually, which some people feel is excessive. On the other hand, Robert Guthrie, a leading proponent of mass screening, argued that

> The detection and treatment of one case of PKU represents an outlay (assuming it is the outcome of 10,000 screening tests, each costing 50¢ to $4.00 in the U.S.) of up to $50,000; but failure to detect that case means a child that must almost certainly be institutionalized for the rest of its life (this assumes an average life span of 50 years and an annual expenditure for custodial care conservatively estimatable at $5,000). The $250,000 figure includes no allowance for the future earnings of the treated case, or of tax income from such earnings. It does not allow for the suffering experienced by the parents of a permanently retarded child... We are not talking here about human values, but purely economic ones: whether it is better to spend $50,000 now or five times that sum later.[2]

By combining the detection of several hereditary defects in one sample from each child, the purely financial returns on the screening costs from detection and euphenic treatment are increased even further.

chemical modification in sickle cell anemia

Several investigators are trying to develop euphenic treatments for sickle cell anemia, which affects an estimated two million people in the world (and three of every thousand African-American children). This tragic condition is characterized by early mortality and crises of extreme pain, when the misshapen sickle cells clog capillaries to interfere with blood circulation. The sickling process is clearly related to the amount of oxygen carried in the blood. In the late

1960s, a group of molecular biologists at Rockefeller University suggested that cyanate, which resembles carbon dioxide, might stabilize sickle cell hemoglobin enough to prevent sickling. Amazingly, cyanate treatment in test tubes of blood cells from sickle cell patients revealed that the chemical binds irreversibly to hemoglobin and inhibits sickling when oxygen is removed, without reducing the oxygen-carrying capacity of the blood cell. However, trials on people showed that cyanate is too toxic for routine use, so alternatives are being investigated.

In controlled trials, hydroxyurea was shown to decrease the incidence of painful crises. Hydroxyurea probably works in multiple ways, including a slight increase in fetal hemoglobin, the hemoglobin of the fetal stage of life, which is not affected by the sickle cell mutation. Other investigators are experimenting with butyrate, which induces fetal hemoglobin synthesis more effectively than hydroxyurea. Since red blood cells sticking to the surfaces of small blood vessels is a critical event in bringing on crises, some investigators have focused on this point. One research group has reported success with polaxamer 188, which reduces the viscosity of the blood and acts on cell surfaces to reduce their adherence. Other groups are using animals to investigate treatments such as antibodies that block the proteins most responsible for cells sticking to one another.

A more extreme euphenic treatment is bone-marrow transplantation, but this carries some risk of mortality, and it is hard to find appropriate donors. The point, however, is that euphenics provides a first line of approaches to treating the symptoms of genetic disease. In later chapters we will discuss gene therapy, a direct modification of the DNA of the abnormal gene itself.

prospects for euphenic intervention

Although several genetic defects can be treated euphenically today, this field is just in its early stages. The number of treatable conditions will undoubtedly grow quickly over the coming years. But the "cured" people who reach maturity and have children will pass defective genes on to their children, increasing the frequency of defective alleles in the population. Will we, as some suggest, become a species of genetically defective individuals dependent on modern science and medicine for the props necessary to correct our hereditary defects?

There is no question that the incidence of some hereditary conditions is increasing. The following cases are not single-gene diseases, but are all genetic diseases with complex, multigenic inheritance. For example, congenital pyloric stenosis is a blockage of the stomach opening to the small intestine that occurs in two to four of every thousand live births. Until the twentieth century, the condition was always lethal in infancy, but in 1912 a simple operation was developed to correct the defect so that affected people can now lead a normal life. When such cured people become parents themselves, about seventy of every thousand of their children have pyloric stenosis, a rate that is twenty times the general incidence. Diabetes mellitus, cleft lip and palate, pyloric stenosis, and congenital heart defects are other relatively common conditions with a complex hereditary basis that are amenable to surgical and medical correction and, hence, can be expected to increase in coming generations. However, as we show in Chapter 15, you should be very wary of exhortations against the imminent deterioration of the human gene pool. The great majority of deleterious recessive alleles reside in normal heterozygous carriers. For this reason, doubling (say) of the frequency of deleterious alleles does not lead to a doubling of the frequency of the actual disorder. The time required to double the incidence of hereditary defects is staggeringly large – hundreds or thousands of years. At the same time, molecular methods are now being developed for identifying heterozygous carriers for many single-gene diseases, and for the component alleles of diseases of complex inheritance. These advances should allow people to voluntarily limit the number of children they have (even to have no children), thus to reduce the frequency of deleterious alleles in a population. Gene therapy techniques, as described in Chapter 12, offer some hope for the distant future, but, on both scientific and humanitarian grounds, euphenic measures are going to be crucial for the treatment of many genetic disorders in the foreseeable future.

The success of euphenics depends on modifying the environment. This raises the broader issue that most disease is not hereditary but is environmentally induced. Furthermore, the expression of multigenic diseases – and even many single-gene diseases – is heavily dependent upon the environment. Hence, as medical geneticist Patricia Baird has pointed out, we must be careful not to oversell a genetic approach to public health. (This point will become more relevant when we look at the great hopes arising out of progress in genomics, in Chapter 12). The biggest bang for the health care buck can still be had at the environmental level.

chapter seven

the hereditary material, dna

From the beginning of scientific studies of heredity, one question remained foremost in the minds of investigators: just what *is* the hereditary material? By early in the twentieth century, the Sutton–Boveri hypothesis – that genes are located on chromosomes – was well established. But what is the chemical substance in the chromosomes that carries the genetic information? Even when biochemistry was quite young, investigators knew that cells contain at least two complex substances – proteins and nucleic acids – that were likely candidates for the honor. Though little was known about the structure of either, protein appeared to be the more complex, and it was generally assumed that genes are made of protein. However, there was other evidence that nucleic acids were involved. When E.B. Wilson wrote his classic book *The Cell in Heredity and Development*, he concluded in one edition that protein was the important material, in another that nucleic acid was. But no one knew for sure.

The answer to this key question emerged through the study of bacteria and of the viruses that grow in them. In a short time, in 1952–3, the whole answer emerged: the stuff of heredity is deoxyribonucleic acid (DNA), and its physical structure accounts for all the major phenomena of heredity. The identification and characterization of DNA as the hereditary material was one of the greatest scientific breakthroughs of the twentieth century. Its implications are staggering. The structure of the DNA molecules in our cells determines a large proportion of our physical features, including the structures of our nervous systems, so one might speculate that an important part of our personalities and behavior is genetically hard-wired. Understanding DNA offers us an important insight into human nature. The story of how we achieved our present understanding and how all the clues were pieced together is an exciting example of scientific detective work and ingenuity.

bacteria

You will recall that bacteria are distinguished from other organisms because they are procaryotic and have no membrane-bound nuclei, in contrast to eucaryotic organisms, including plants and animals, whose cells have real nuclei. Furthermore, they are so small that the 1000 × magnification of a good microscope is needed to see them clearly, and even then their structure is only shown well by an electron microscope. Figure 7.1 shows the relative sizes of some common bacteria along with several viruses, which are still tinier than bacteria and have also played a major role in genetic work.

Although we tend to associate bacteria with disease, the overwhelming majority of them are benign creatures living in natural waters, soil, or other organisms. The best-studied bacterium is *Escherichia coli* (*E. coli*),[1] one of several kinds of bacteria that inhabit the human colon (large intestine) and make up much of the weight of feces. (These bacteria help keep our digestive systems healthy and also provide us with some vitamins.) We study bacteria partly to understand and control pathogenic (disease-producing) types, but they are useful tools for pure research because they are relatively simple, single-celled organisms that grow rapidly in simple nutrient growth media. An organism such as *E. coli* grows in a solution containing a sugar, which supplies energy and carbon atoms, plus several salts such as magnesium sulfate and ammonium chloride that provide the other elements they need. After sterilizing a flask of such a growth medium by heating, we inoculate it by carefully introducing a few cells of a bacterium such as *E. coli*. Each cell assimilates materials from the medium and converts them into more of its own substance, using its enormous stock of enzymes, so it grows larger. (We should emphasize that much of what the cell must make as it grows – much of its mass – is just more of those enzymes.) After a short time, perhaps only thirty minutes, it divides

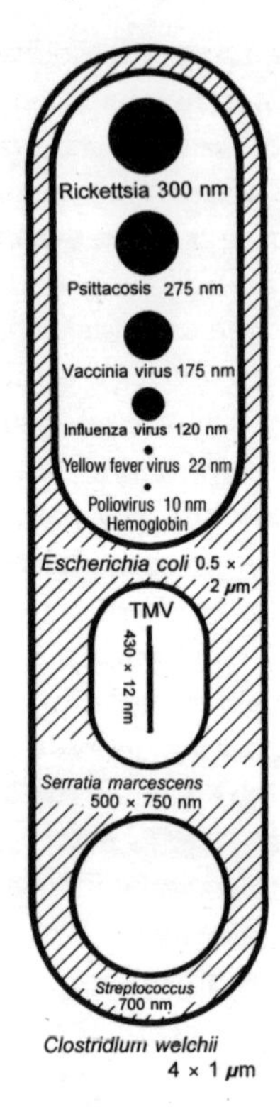

Fig. 7.1. *A large bacterium,* Clostridium, *is about 1 × 4 μm (micrometers). 1000 μm = 1 mm; 1000 nm (nanometers) = 1 μm. Smaller bacteria, such as* E. coli *and* Serratia, *could fit inside, along with still smaller bacteria and viruses; TMV is tobacco mosaic virus.*

in half. Each of the resulting two cells continues to grow and divides to make four; the four make eight, the eight sixteen, and so on. This pattern is called *exponential growth*; it is the same as the growth of money invested at compound interest, because each cell continues to make more cells, just as each dollar earned as interest is added to the capital to earn still more. (With bacteria, however, the interest rate may be a whopping one hundred percent per half hour.) A few bacteria can grow to enormous numbers in a short time. They stop only when they have exhausted some nutrient in the medium or cannot get enough oxygen, or when too much of their wastes accumulates.

A culture medium containing *agar* (an inert thickening substance from seaweed, which is also used to give body to ice cream) becomes semi-solid like gelatin, and this solid medium can be used to culture and examine bacteria. When poured as a hot, sterile liquid into shallow dishes called petri plates, the agar medium hardens into a solid layer as it cools. We can easily place a small sample of bacteria on the surface of the agar with pipettes or sterile wire needles, and spread them around with a sterile glass rod. This operation is called *plating* the bacteria. Each cell grows where it lands on the plate, so all the daughter cells of an original cell remain clustered in one spot. Eventually enough cells accumulate at each spot to form a visible *colony* (Figure 7.2). Colonies have distinctive colors and shapes; we can use them to identify bacteria and study their characteristics. Each colony is also a *clone*. This term – which has some special meanings in modern molecular genetics – just means a group of individuals or cells produced by asexual cell division. Strawberry plants propagate by sending out runners which put down new roots, and all the plants that arise in this way are also a clone. Similarly, isolated plant and animal cells of many kinds can be propagated in culture flasks with nutrient media, where they grow and divide for a long time. The population of cells derived from a single cell in this way also constitutes a clone.

the first clue

In 1928, Frederick Griffith demonstrated that a substance from dead cells of one strain of bacteria can transfer their characteristics to live

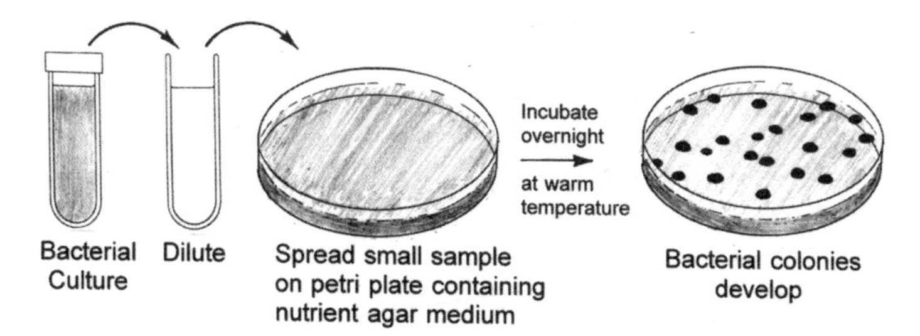

Fig. 7.2. *If a culture of bacteria is diluted and a sample is spread on the surface of a petri plate, each cell grows into a visible colony, a single clone of bacteria.*

cells of another strain. He knew that strain IIIS of *Diplococcus pneumoniae* could cause a fatal pneumonia in mice, whereas strain IIR cells were relatively harmless. He then heat-killed some IIIS cells, mixed the debris with live IIR cells, and injected mice with the mixture. These mice also died. The live cells had picked up some material from the dead cells that *transformed* them and gave them IIIS characteristics. In 1944, Oswald T. Avery and his colleagues at the Rockefeller Center, in New York, showed that the transforming material is DNA. They could destroy proteins and other materials in the cells without affecting transformation, but any procedure that destroyed DNA eliminated transformation.

This was the first solid piece of evidence that hereditary information could be identified with a specific material and that the material was a nucleic acid. However, this work did not impress many people. It did not fit other information available at the time indicating that DNA is quite a simple molecule that could not be the stuff that genes are made of; most biologists expected proteins to be the molecules of heredity, and their preconception colored their ability to assess the data objectively. More convincing evidence came in 1952 from a classic experiment of a different kind.

VIRUSES

In contrast to organisms, which always consist of one or more cells, there are subcellular objects that do not have a cellular structure and can reproduce themselves only by penetrating a living cell. They are called viruses.

The ancient Romans recognized that certain diseases seemed to come from other animals, so they called such a "poison" a virus, and diseases caused by them were said to be virulent. Of course, the Romans couldn't see such poisons, and by their criteria chemical poisoning (for example, botulism), bacterial diseases (typhoid), and virus infections (polio) were all alike. By the Renaissance, two classes of diseases were recognized, contagious and non-contagious, and the term "virus" was reserved for the former. In the nineteenth century, after the work of Pasteur, microscopes enabled microbiologists to see bacteria, which became known as the "viruses of disease." By the twentieth century, it was known that many causative agents of disease could not be cultured and were so small that they could pass through filters that would trap all known bacteria. So the term "virus" was reserved for these filterable agents. With the advent of the electron microscope, viruses can now be seen in the full complexity of their structures.

In 1915, Frederick Twort reported that plates of bacteria (*Micrococcus*) often became watery or glassy. Such areas no longer contained live bacteria, but did have a killing factor that could spread to other bacteria. Twort's discovery elicited little interest, but in 1917 Felix d'Herelle reported on "an invisible microbial antagonist of dysentery bacilli." D'Herelle later reported that in 1910, while he was studying bacteria that cause diarrhea in Mexican locusts, he had begun to suspect the existence of a "disease" of the bacteria. He had found that in a thick spread of the bacteria he would get clear spots where all the bacteria were dead. This observation had convinced him that bacterial viruses could be a tremendous weapon to fight disease. In 1915, he set out to find viruses that could kill *Shigella*, the dysentery organism. He described his discovery of such a virus:

> The next morning, on opening the incubator, I experienced one of those rare moments of intense emotion which reward the research worker for all his pains: at the first glance, I saw the broth culture, which the night before had been very turbid, was perfectly clear: all the bacteria had vanished, they had dissolved away like sugar in water. As for the agar spread, it was devoid of all growth and what caused my emotion was that in a flash I had understood: what caused my clear spots was in fact an invisible microbe, a filterable microbe, a filterable virus, but a virus parasitic on bacteria.
>
> Another thought came to me also: "If this is true, the same thing has probably occurred during the night in the sick man, who yesterday was in a serious condition. In his intestine, as in my test tube, the dysentery bacilli will have dissolved away under the action of their parasite. He should now be cured." I dashed to the hospital. In fact, during the night, his general condition had greatly improved and convalescence was beginning.[2]

D'Herelle called the bacterial viruses *bacteriophages* (eaters of bacteria). His ecstatic hopes for a potent anti-bacterial agent to

(continued overleaf)

(continued)

cure disease led to the isolation of phages specific to anthrax, bronchitis, diarrhea, scarlet fever, typhus, cholera, diphtheria, gonorrhea, bubonic plague, and osteomyelitis. In the enthusiasm over the potential for phage therapy, the author Sinclair Lewis created medical scientist Martin Arrowsmith who discovers an "X-principle" that is the same as d'Herelle's bacteriophage.

For many years, the therapeutic potentials of phages were ignored in most countries because antibiotics, which came into general use during the Second World War, were so powerful and provided such easy cures of bacterial diseases. However, phage therapy continued to be important in Eastern Europe, especially in Georgia, Poland, and the Soviet Union. In the last few years, particularly because so many common pathogenic bacteria are becoming resistant to all the antibiotics that have been used against them, phage therapy is again coming into vogue. But that is another long story that we cannot tell here.

bacteriophages

About 1915, the Englishman Frederick Twort and the Canadian Felix d'Herelle independently discovered agents called *bacteriophages* that produced infections of bacteria (see Box, pp. 112–14). The very idea may sound bizarre to us, since we generally think of bacteria themselves as the causes of infections; but it is a maxim of parasitology that "bigger fleas have lesser fleas upon their backs to bite 'em," and even a very small creature can have a smaller one that lives upon it. Bacteriophages – or, as we now generally say, *phages* – are viruses that grow in bacterial cells, just as many other kinds of viruses grow in plant or animal cells. They are good examples of how viruses in general behave.

Viruses, in the first place, are not organisms. Each organism consists of one or more cells, but viruses are minute particles called *virions*, much smaller than the smallest cell, that are little more than a piece of nucleic acid (either DNA or RNA) wrapped in a protective covering of protein (Figure 7.3). Most virions are either small spheres or rods. Animal viruses usually get into their host cells by

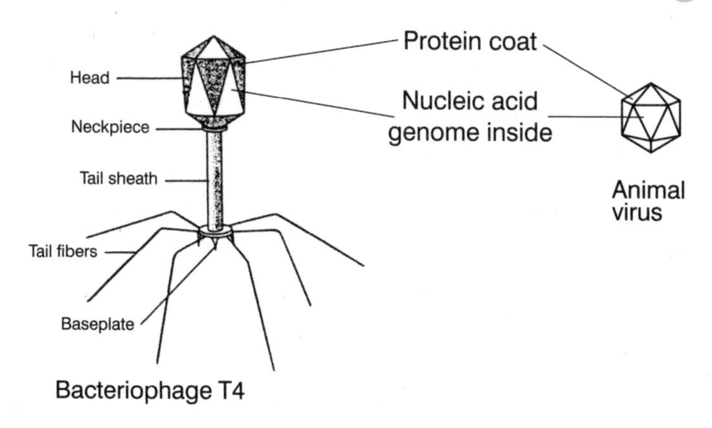

Fig. 7.3. *The general structure of virus particles, or virions, of a bacterial virus and an animal virus.*

attaching to the cell surface and being engulfed (as if they were food particles); plant viruses usually get into their host cells through wounds created by insects or worms. Many phages, like those shown in the figure, attach to the bacterial cell surface by their tails. This could be seen even in the earliest electron micrographs made around 1945. It was also known that phages are about half protein and half DNA. Although little more was known about them, they promised to be excellent tools for genetic studies.

Phages are easy to work with because they multiply rapidly and produce enormous numbers in a short time – 100–200 new phages in about half an hour. They are also easily grown on petri plates (Figure 7.4); a few phages are mixed in warm, melted agar with some bacteria and the mixture is poured over a layer of nutrient agar in a plate. The bacteria grow in a thin, uniform layer, called a *lawn*, in the surface of the agar, and each phage makes a clear spot, or *plaque*, where it initiates a local infection and kills the cells, making a hole in the bacterial lawn. We can determine the number of phages in any material by counting the plaques produced by a small sample of it. Furthermore, the size and form of a plaque is genetically determined and is characteristic of each type of phage.

Careful studies of phage multiplication were initiated in the early 1940s by Max Delbrück, Salvador Luria, and Alfred D. Hershey, who formed what was informally known as the American Phage Group.

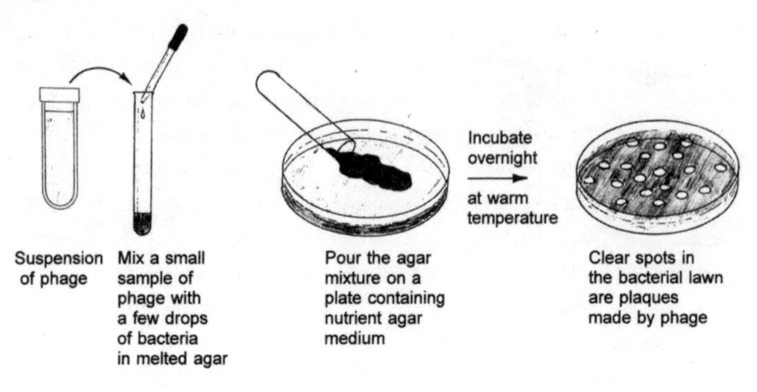

Fig. 7.4. *Phage are grown on petri plates by mixing a sample of phages and bacteria in which they can multiply in a tube of melted agar. This mixture is poured over the agar in a plate; when the bacteria grow up after several hours, they form a uniform lawn. Each of the clear, round holes in the lawn is a plaque, a spot where the progeny of one phage particle have infected and killed the bacteria.*

The details of phage multiplication have been elucidated by their students and followers, concentrating on a few types of phage designated T1 through T7 that grow in *E. coli*. Our present general understanding of gene structure and function has emerged largely from studies of phages and bacteria.

the hershey–chase experiment

Knowing that phages are made of DNA and protein, in about equal amounts, Alfred Hershey and Martha Chase set out in 1952 to identify the functions of the two components in phage T2 by *labeling* them, which means incorporating radioactive atoms into these molecules so their fates could be traced. They knew that proteins contain sulfur but no phosphorus whereas DNA is rich in phosphorus but contains no sulfur. So they grew one batch of phages with radioactive sulfur (^{35}S), which was incorporated into their proteins as a label, and another batch with radioactive phosphorus (^{32}P), to label their DNA. Then each batch was used to infect bacteria.

Electron microscopy had shown that at least part of the lollipop-like virions stay attached on the outside of phage-infected cells. So

Hershey and Chase put the infected cells in a blender and whipped them around to shear off these virions. Then they centrifuged the bacteria to the bottom of a tube and measured the radioactivity in them and in the supernatant, the liquid that does not pellet to the bottom. Cells infected with ^{32}P-labeled phage were highly radioactive while those infected with ^{35}S-labeled phage were not, and much of the radioactive sulfur was in the supernatant after the cells were blended. This showed that the DNA had been transferred into the cells, where it could not be removed, while the remainders of virions on the cell surface, which could be sheared off, were made of protein.

Hershey and Chase then showed that the progeny phages made by these cells contained much of the labeled DNA, but they had little or none of the protein. Since the material injected into the cells directs the production of more phages, the inescapable conclusion was that *DNA is the genetic material itself.* In retrospect, we can see that in bacterial transformation, live cells must pick up some DNA extracted from other cells and replace some of their own genes with this foreign DNA.

The Hershey–Chase experiment has an important implication that we mustn't miss. Phages are made entirely of DNA and protein. Only DNA goes into the cell to initiate an infection; half an hour later, new phages come out that are made of DNA and protein again. Therefore, *the function of the DNA must be to carry the information for making those proteins.*

A multitude of experiments with phages have now shown that phages multiply as shown in Figure 7.5. When a phage such as T2 or T4 attaches to the cell surface by its tail, the phage DNA is transferred into the host cell. Within minutes, the phage DNA starts to direct the synthesis of new phage proteins. First, several proteins are made that turn off host functions; some of them stop synthesis of the host's proteins and others are enzymes that begin to destroy the host DNA. Another series of enzymes start to replicate the phage DNA. The synthesis of these enzymes soon stops. Then new genes are turned on that specify the many structural proteins of the phage *capsid*, its tadpole-shaped protein coat. These capsid proteins spontaneously assemble themselves into new pieces of capsid – tail pieces, heads, and tail fibers. Other enzymes pack the new heads with new pieces of phage DNA. By about thirty minutes after infection, the cell is typically filled with a couple of hundred new virions, and it generally

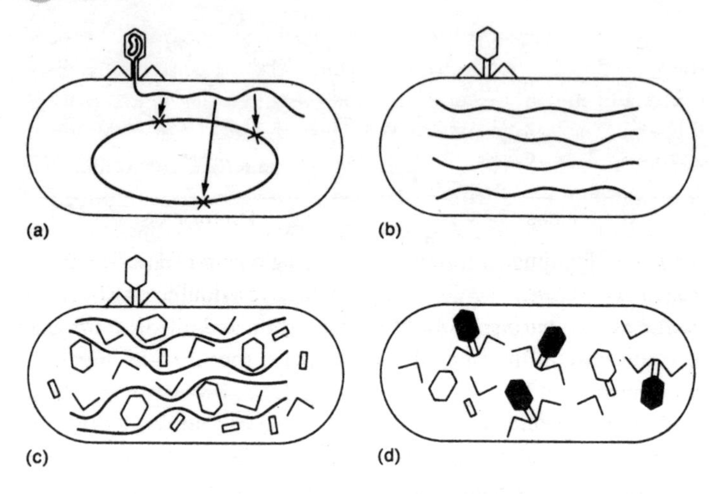

Fig. 7.5. *A summary of the process of phage infection. (a) The infecting phage injects its DNA into the cell, and this DNA begins to change the cell into a factory for making more phages. Enzymes encoded by the phage DNA attack the host DNA and stop the expression of host genes. (b) The phage DNA is replicated many times, and new phage capsid proteins are synthesized (c). The new capsid proteins combine with the DNA to make new phage particles (d), which are released when the cell bursts (lyses).*

then disintegrates, or *lyses*, under the influence of other phage enzymes.

dna structure

Remember that the main components of an organism are polymers. Nucleic acids are polymers quite different from proteins; they are also called *polynucleotides*, because their monomers are *nucleotides*. A nucleotide has three parts: a *base* linked to a *sugar* which is linked to a *phosphate* (PO_4). A nucleic acid is named for its sugar; ribonucleic acid (RNA) contains ribose, and deoxyribonucleic acid (DNA) contains deoxyribose (ribose with one oxygen atom removed). The bases are large, ring-shaped, nitrogenous molecules; DNA nucleotides have one of four bases: *adenine*, *guanine*, *cytosine*, and *thymine* (A, G, C, and T; in RNA, *uracil* (U) replaces thymine).

Adenine Guanine Thymine Cytosine Uracil

Purines Pyrimidines

Cytosine, thymine, and uracil have a single ring of atoms and are called *pyrimidines*; adenine and guanine have a double ring of atoms and are called *purines*. Notice that the carbon and nitrogen atoms in the rings are all numbered for reference; those of the sugar are numbered from 1′ to 5′.

A polynucleotide (either DNA or RNA) is made by linking the phosphate of one nucleotide to the sugar of another – specifically, from carbon atom 3′ on one nucleotide to carbon 5′ on the next:

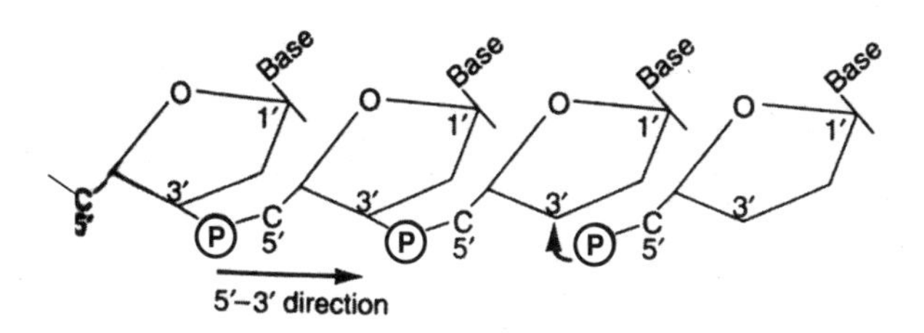

This gives every DNA chain a 3′→5′ *polarity* just as a protein chain has a polarity from the amino end to the carboxyl end. The bases themselves stick off to the side of the sugar–phosphate backbone of the molecule.

Until about 1952, people generally thought that DNA consisted of the four kinds of nucleotides repeated again and again in a regular sequence, so all DNA molecules were more or less the same and they could not carry information. But when Erwin Chargaff carefully analyzed the composition of DNA from various organisms, he found that the nucleotides were not present in equal amounts, and that:

1. the total of purines (A + G) always equals the total of pyrimidines (C + T);
2. the amount of A always equals the amount of T, and the amount of G always equals that of C (A = T, G = C);
3. the ratio of (A + T) : (G + C) varies considerably from one species of organism to another.

In 1953 James Watson and Francis Crick, at Cambridge University, finally determined the structure of DNA. Watson, being a student of Luria's, was a member of the Phage Group and knew about Hershey and Chase's work. Crick was a physicist who knew the potential of a powerful analytical method, *X-ray diffraction*. X-rays can be used to reveal the structure of molecules, even though we can't focus them as we can focus light beams with a microscope. X-rays are aimed at a crystalline material, and the atoms of the crystal scatter them in a predictable way onto photographic film; we can interpret the pattern they make to show how the atoms are arranged in the crystal. Using this technique, Maurice Wilkins and Rosalind Franklin, at University College, London, found that DNA apparently has a helical structure – the shape of a corkscrew. Watson and Crick tried to develop a model for DNA structure, using atomic models of nucleotides. They were successful because they insightfully combined information from Wilkins and Franklin's work with Chargaff's data and their general knowledge of DNA's role in heredity. Watson tells this story in his autobiographical *The Double Helix*, a book that should be read along with Anne Sayre's *Rosalind Franklin and DNA* for a balanced view of Franklin's work.

Watson and Crick's significant insight was that the specific bases were all-important and that Chargaff's rule "A = T, G = C" meant that the bases must be paired in those combinations. They proposed that a DNA molecule consists of two polynucleotide strands in a helical arrangement, with their 3′–5′ polarities running in opposite directions (Figure 7.6). The strands are held together in the middle by the bases from one strand bonding to the bases from the other, but adenine can bond only to thymine and guanine only to cytosine:

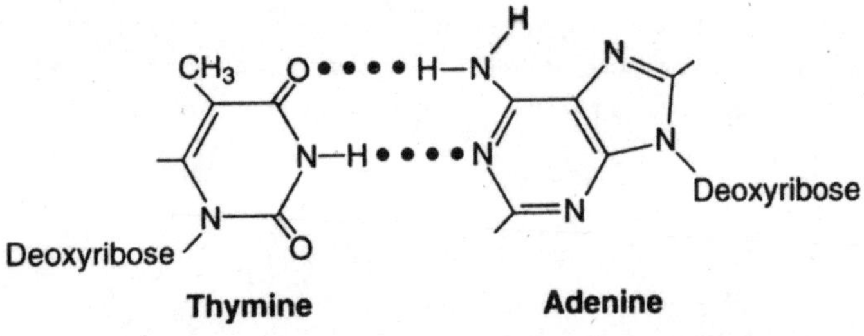

Deoxyribose Deoxyribose

Cytosine Guanine

The bases are held together by weak attractions called *hydrogen bonds*, in which two atoms that have a little excess negative charge, such as O and N, are held together by a hydrogen atom (that has a little excess positive charge). We say that bases that can bond this way are *complementary* to each other, in the way that a hand is complementary to a glove or that the curve of one jigsaw piece is

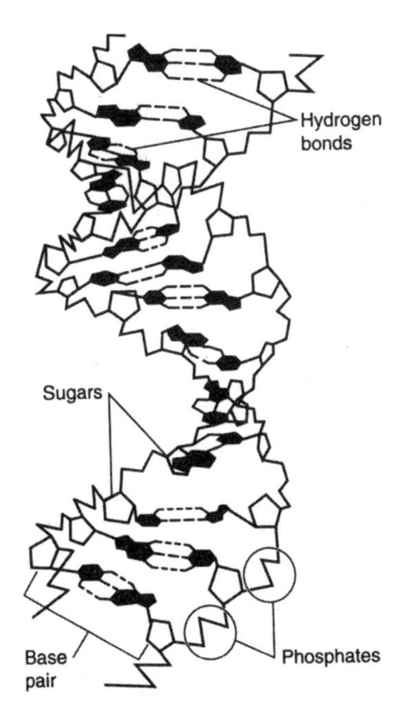

Fig. 7.6. *The form of the DNA double helix, as proposed by Watson and Crick.*

complementary to the curve of the piece it fits with. Complementary base pairing accounts for virtually everything in heredity, and for much of biology in general. It accounts, obviously, for Chargaff's rules; it shows how the two strands stay together; and, as we will now show, it accounts for the most essential genetic properties of DNA. Watson and Crick's short note about their model in the journal *Nature* in 1953 hardly hinted at the revolution that would result.

genetic implications

Unlike Mendel's work, Watson and Crick's paper was instantly recognized by the scientific community as a monumental contribution because it provided a clear basis for inheritance. First, we can see at once that the sequence of bases in each DNA molecule can carry all the information of heredity, using some kind of *genetic code*. Information is commonly carried in sequences, such as the sequences of letters and punctuation marks in writing or the sequences of dots and dashes in Morse code. Furthermore, that information must be transmissible from one DNA molecule to its daughter molecules as the genetic material is duplicated with each generation and passed on to the next generation. The process of making two exact replicas of one original DNA molecule is called *replication*, and the Watson–Crick model shows how this can occur.

In a DNA molecule, each nucleotide has its complementary nucleotide; each strand as a whole is the complement of the other. Replication is carried out by a complex enzyme, *DNA polymerase*, which starts to unwind the double-stranded molecule to leave exposed unpaired bases on each arm, rather like opening a miniature zipper (Figure 7.7). The description we are about to give is highly simplified, since the biochemistry of replication is complicated, especially because new DNA strands can grow only at their 3′ ends. The essence of the process is that DNA polymerase molecules move along each single strand, synthesizing complementary strands to make two double helices where before there was one. Each exposed base can pair with its complementary nucleotide; wherever a C is exposed, it attracts a new G; the opposite strand has an exposed G, which attracts a new C, and similarly for A and T. The cell has lots of free nucleotides because its metabolic apparatus is constantly

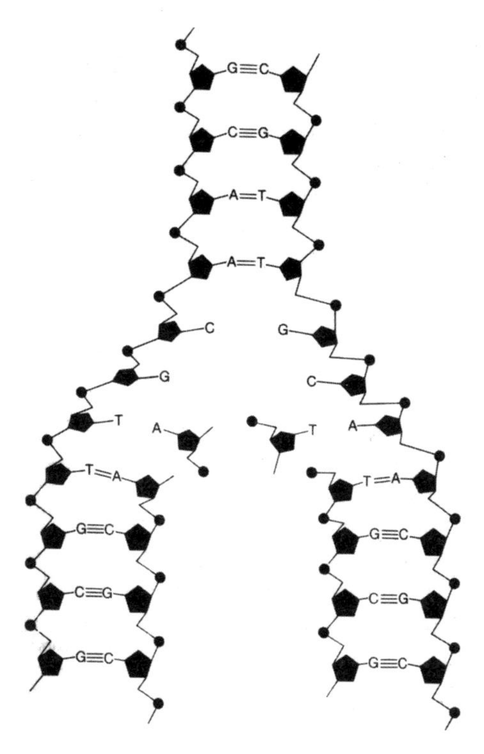

Fig. 7.7. *In DNA replication, an enzyme complex unzips the double-stranded molecule and each of the exposed bases attracts a new complementary nucleotide. This process continues on both strands until two identical double-stranded molecules have been made.*

churning them out, and the polymerase links them one by one into new strands. Thus each strand dictates the formation of a complementary strand identical in base sequence to its former partner, all because of the specificity of base pairing, eventually creating two double helices identical to each other and to the original molecule.

The nucleotide sequence of DNA must store genetic information, and the final implication of the Watson–Crick model is that a *mutation* could result from any change in sequence, either because one base pair is substituted for another or because the sequence is broken and rearranged. Such events rarely occur, and even when they do a cell has mechanisms for repairing and correcting some faults. Nevertheless, each organism has an enormous amount of DNA, and even if the probability of inserting the wrong base is only one in a million, an average of ten errors will be made in every ten million bases, so mutation is a force to be reckoned with. We explore the nature of mutations in more detail in Chapter 14.

testing dna structure

The real scientific value of any model or theory is in its having implications that can be tested. The Watson–Crick model not only incorporated all the known facts about DNA and explained the facts of heredity in outline, but also encouraged a number of predictions.

Look at DNA replication from a broader viewpoint. Each strand stays intact, but they separate from each other and a complementary new strand is made on each old one. This is called *semiconservative* replication, because the whole double helix is not conserved but each of its strands is. Suppose the original molecule were colored red and newly polymerized nucleotides were colored green. Then after one round of replication, each daughter molecule would be half red and half green. If they replicated again, two of the molecules would still be half red and half green and the other two would be all green. Representing the original molecule by solid lines and new molecules by dotted lines, we have:

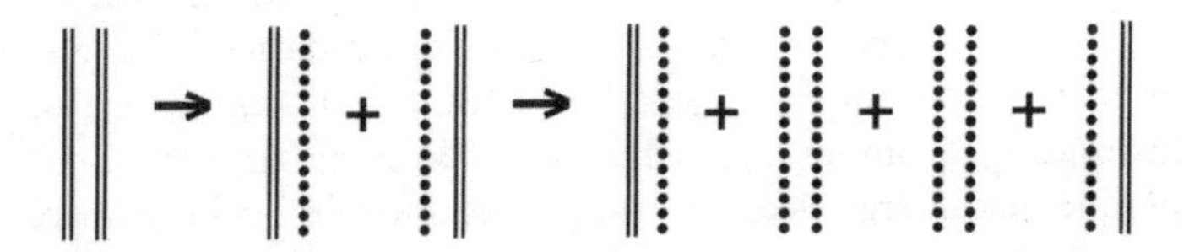

In 1957, Matthew Meselson and Franklin Stahl, at Caltech, found a way to test this idea. Working with Jerome Vinograd, they discovered that a dense solution of the salt cesium chloride (CsCl) will form a *density gradient* when it is centrifuged rapidly. An instrument called an *ultracentrifuge* can be made to spin materials at almost sixty thousand revolutions per minute; that is, one thousand revolutions per second. A little experience with the relatively slow rides at an amusement park will show you how much force a spinning object can exert, and an ultracentrifuge exerts such enormous force that it can push the rather heavy cesium ions in a solution toward the bottom (outside) of a chamber. After spinning for many hours, a CsCl solution forms a gradient with slightly higher density toward the outside and lower density toward the inside. DNA molecules in such a solution will come to rest at the point where their density is equal to that of the solution.

Meselson and Stahl grew bacteria in a medium containing heavy nitrogen atoms (^{15}N, as opposed to ordinary ^{14}N). The cells incorporated these atoms into the DNA they made, thus making it all more dense than normal. (This effectively painted the DNA "red.") Then they transferred the cells to a medium containing only ordinary nitrogen, so the DNA made after that time was all light (that is, "green"). They took samples at various times and determined their density by spinning them in CsCl. (The ultracentrifuge has an optical system and camera for locating the DNA.) The DNA was initially all dense. After one round of replication, it all became half dense. After a second round of replication, half of it was half dense and half was light. This is exactly what they should have found if DNA replicates according to the Watson–Crick model.

A second implication of the model is that one should be able to see a fork in replicating DNA. The two strands cannot separate all at once; they should start to separate at one end, and new strands should be made on these separate single-stranded regions. DNA molecules can be visualized by the technique of *autoradiography*, which depends upon incorporating atoms of a radioactive isotope, or *radioisotope*, into the molecule. Tritium (^{3}H), an isotope of hydrogen, is commonly used because decaying tritium atoms produce low-energy electrons that do not travel far before they are absorbed by some matter. If such an electron is absorbed by photographic film, it makes a black spot, showing where the tritium atom was; a picture that results from the decay of many atoms is called an autoradiograph because the material takes its own picture through radioactivity.

An autoradiograph of DNA is made by growing some cells (such as bacteria or rapidly growing plant roots) in a medium containing thymidine (one of the DNA nucleotides) that has been labeled with tritium. This is incorporated into any DNA that is being made. The material is spread in some appropriate way on a slide – a plant root tip, for instance, may simply be squashed carefully. After unincorporated thymidine is washed off, the slide is covered with photographic emulsion and sealed in a dark box, sometimes for months. When the emulsion is developed, just like a film, black silver grains will appear wherever a tritium atom decayed. The cells may be stained to make them visible with a microscope.

Autoradiographs have been made by labeling *E. coli* chromosomes, then gently extracting them from their cells and floating them on a water surface so they spread out. This technique shows that the molecule is a ring, over a millimeter in total circumference – about a thousand times as long as the cell it is packed in. The chromosome is generally caught in the midst of replication, and, from this and other work, we can see that replication actually occurs at two forks, with two DNA polymerases moving in opposite directions. Thus, although the Watson–Crick model has been strongly supported by experimental evidence, research has provided its share of surprises as well.

The single *E. coli* chromosome is the organism's entire genome. It is a comparatively small genome, yet it contains about 3.8×10^6 nucleotide pairs. Molecular biologists are a bit sloppy in describing the sizes of DNA molecules; the proper unit is a nucleotide pair (n.p.), but we tend to equate a nucleotide with a base and measure lengths in "bases" or more commonly in *kilobases* (kb) – that is, in units of a thousand bases or nucleotide pairs.

chapter eight

the genetic dissection of gene structure

We have now developed the concept that a gene is a segment of DNA located on a chromosome, and that it specifies the structure of a polypeptide chain. Mutations can occur in a gene, and it is primarily through these mutations that we know the gene exists. Because mutations produce defects, such as the loss of an enzyme, we know that the normal gene and the normal enzyme exist, and that one carries the information for the other. There remain unanswered questions about how genes are constructed and how they work. In this chapter, we will show how genetic analysis can be used to show the structure of genes in more detail and to analyze the relationship between a gene and its protein product.

gene arrangement

Our knowledge that genes are on chromosomes raises a problem: human cells have only twenty-three pairs of chromosomes, yet thousands of different hereditary conditions are known in humans and therefore, presumably, thousands of different genes. The number of X-linked traits alone is already in the hundreds, and even the shortest autosome carries hundreds of genes. What does this do to the Mendelian notion that different genes assort independently? It means that the law of independent assortment applies only to genes that are on different chromosomes; it was necessary to recognize this law for the simplest situations, to make sense of basic patterns of inheritance. In fact, many genes are on the same chromosome, so they tend to be inherited together. We say they are *linked*. One triumph of modern genetics has been the ability to develop *linkage maps* for many organisms, showing the locations of genes on their chromosomes, and we will see that this has not only theoretical but also quite practical uses.

The place a gene occupies on its chromosome is its *locus* (plural, *loci*). Except for certain rare events that produce rearrangements of chromosomes, the locus of a gene is the same across all members of the species. We have already indicated that a gene is really known by mutations that occur in it, which generally make it defective or unusual in some way. Most hereditary features are known only because of obvious deficiencies such as hemophilia, color-blindness, and phenylketonuria. The normal allele of a gene is called the *wild-type*, although the term is useful only for certain experimental organisms: for ordinary human characteristics, such as eye color or blood type, no one allele can be considered the wild-type. And wild populations carry several alleles for many genes. A mutant allele can be used as a *marker* that helps us locate the gene. Thus, we can use mutations in the genes for human hemoglobin that produce such conditions as sickle cell anemia as markers to find the loci of those genes. Without such variant forms of a gene, we would have little ability to study it.

The genetic map of a chromosome is just a line with gene loci placed at their proper relative distances from one another, measured in *map units* (m.u.). Although some microscopic methods allow us to identify a gene locus with a place one can actually see on a chromosome, usually we can only map one locus relative to another. To do this, we create some individuals that are heterozygous for two genes so the two marked chromosomes can interact with each other. The arrangement of alleles in those individuals is called the *parental* combination; we picture it with gene symbols along two lines representing chromosomes, and we generally simplify this to a single line representing both chromosomes:

$$\frac{A \quad B}{\overline{a \quad b}} \quad \text{becomes} \quad \frac{A \quad B}{a \quad b}$$

The markers above the line are on one homolog and those below are on the other. For convenience of printing, the line may be shown as a diagonal, so this arrangement is *A B*/*a b*.

To illustrate the principles of mapping, consider two genes that are known to be linked on the X chromosome: color-blindness and hemophilia, where *c* is the allele for color-blindness, *C* for normal vision, *h* for hemophilia, and *H* for normal blood clotting. Since we can't go around mating people as we choose, we have to collect data involving women who are heterozygous for both genes. The markers can be arranged in two ways: in *coupling*, with the dominant alleles on one chromosome and the recessives on the homolog, or in *repulsion*, with the dominant allele of one gene and the recessive of the other on the same chromosome. First, consider women who carry the alleles in repulsion, *C h*/*c H*, meaning they have one *X* chromosome with the alleles *C* and *h* and the other *X* chromosome with the alleles *c* and *H*. Since sons get their *X* chromosome from their mothers, the phenotypes of all male offspring directly show which chromosome they have received. We then expect half the sons to have color-blindness but no hemophilia (*H c*/Y) and half to have hemophilia but no color-blindness (*h C*/Y). But in fact we find:

$$9\ C\ h/Y : 1\ C\ H/Y : 1\ c\ h/Y : 9\ c\ H/Y$$

Thus, ten percent of the sons, whom we call *recombinants*, have a combination of alleles different from the parental combinations that their mothers have. How can this happen?

In prophase of meiosis, homologs pair up and appear to be held together primarily at the chiasmata, the places where their chromatids twist about each other. Sometimes in a chiasma, the chromatids physically break and then exchange segments, in a process called *crossing over*. If a crossover happens to occur between the loci of the two genes we are studying, their alleles will be rearranged:

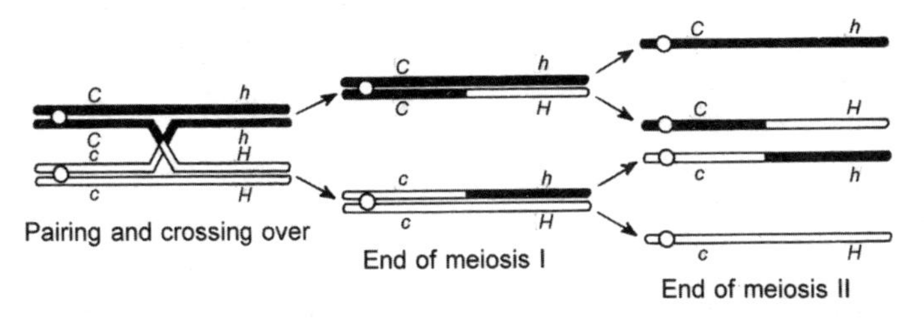

Crossing over produces recombination, and the frequency of recombination, denoted by *R*, is the number of recombinants divided by the total number of offspring; these particular genes recombine with a frequency of ten percent (two recombinants out of twenty), or $R = 0.1$.

Since crossovers occur at random, R depends on the probability that a crossover will occur in the short distance between the genes. Think of this distance as a kind of target that can be hit by accidental crossovers; if the distance is very short, there is little chance that a crossover will occur there and R will be small, but if the distance is larger, crossovers will fall there more often, creating more recombinants. Thus, R is a measure of distance along the chromosome. We arbitrarily call one percent recombination one map unit, so the ten percent recombination between *C* and *H* means that their loci are ten map units apart. We can show that the frequency of recombination is determined only by the distance, and not the original arrangement of alleles, by examining other cases in which the alleles are arranged in coupling, with these results:

9 C H/Y : 1 C h/Y : 1 c H/Y : 9 c h/Y

This is exactly what we expect: ninety percent of the original types and ten percent of the recombinants.

In using humans to illustrate mapping, we have been doing crosses in a rather odd sense. In most organisms, where we perform matings of our choice, mapping experiments require two stages: first, a cross between homozygotes carrying the desired alleles to produce heterozygotes where recombination can occur, and then a second generation to examine the offspring of those heterozygotes. In humans, the first stage has already been performed through marriages that we could not choose, and we are merely examining the results of the second stage.

Having placed two genes relative to each other, we can add other genes, one at a time. For instance, there may be a gene, marked by alleles *A* and *a*, that is linked to *C*. We can then examine sons of women with the genotype *A c*/*a C*, with the results:

43 A c/Y, 7 A C/Y, 8 a c/Y, 42 a C/Y

Here we find fifteen recombinants (7 + 8) out of a hundred, or fifteen percent recombination, so we want to place the *A* gene fifteen map units from *C*. However, the three genes could lie in the sequence *H C*

A, in which case the distance between *A* and *H* would be about 10 + 15 = 25 units; or they could lie in the sequence *H A C*, so *A* and *H* would be only 15 – 10 = 5 units apart. A third cross, involving the *A* and *H* genes, will determine this number.

Sex-linked genes are relatively easy to map, since the gene arrangement in at least one of a woman's X chromosomes can be determined from her father's X, and her sons show their X chromosome genotypes directly. But mapping autosomal genes in humans is difficult. Excellent gene maps (Figure 8.1) have now been created for several laboratory and agricultural plants and animals, but in humans it is hard to determine the exact genotype of individuals or whether alleles are linked in repulsion or coupling.

Linkage maps of human chromosomes are of immediate value in aiding genetic counsellors to make predictions. For example, an

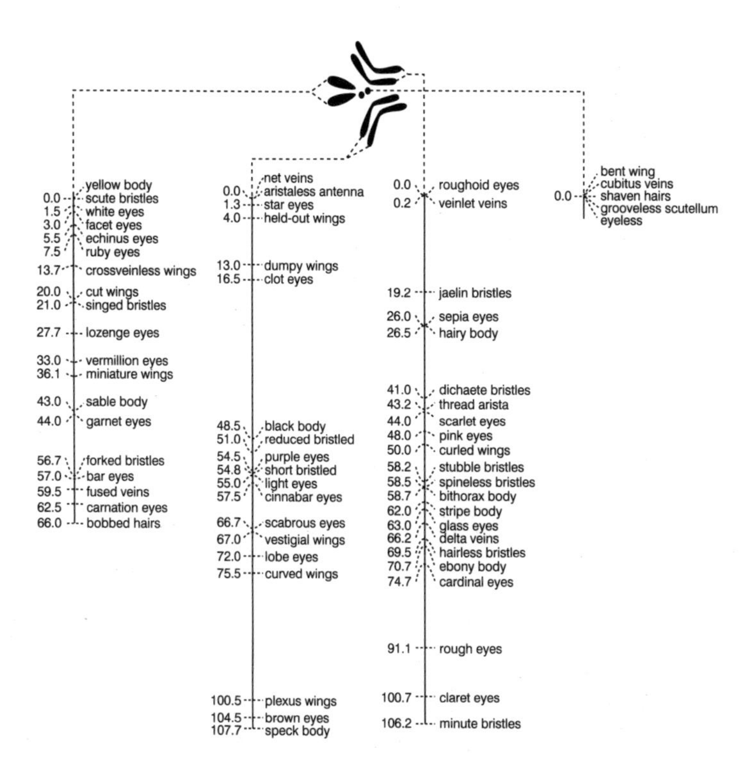

Fig. 8.1. *A genetic map of the fruit fly* Drosophila melanogaster.

autosomal dominant mutation, *Ht*, produces Huntington's disease, a condition marked by degeneration of nervous tissue, usually manifested after middle age. A child of someone with Huntington's disease has a fifty percent chance of carrying the *Ht* allele. But could the child's chances of carrying *Ht* be narrowed down? Suppose there is another gene with two alleles, *A* or *a*, five map units from *Ht*, and that it is easy to determine which allele someone carries. Suppose, also, we know that the affected parent has the genotype *A Ht*/*a ht* and the normal parent is *ht a*/*ht a*. Then we can make some predictions. A child who is homozygous for *a* then has a ninety-five percent chance of being free of *Ht*, since *Ht* is linked to *a* in the affected parent and will be separated from it only five percent of the time. Similarly, a child carrying *A* will also carry *Ht* ninety-five percent of the time. Someone considering having children would benefit from knowing that he or she carried the *Ht* allele as yet unexpressed.

This raises the agonizing question of what value there is in knowing that one carries a gene that will inevitably lead to incurable physical or mental debilitation and/or death. Many would choose to not know, and in a free society that is presumably their right. (The issue of having the right to not know something is a new one, engendered by the accumulated knowledge of modern science.) On the other hand, some people will welcome such foreknowledge; it could relieve a person of an enormous Damoclean sword by altering the statistical odds, and make it easier to make realistic plans about having children. Furthermore, as medical knowledge improves and early treatments develop for conditions that usually appear later in life, it will become more and more valuable to know early on that one carries an allele of this kind.

The allele we have called *a* might be of a gene with recognized function and phenotype, or it could be a site of neutral DNA variation such as a restriction fragment length polymorphism (see below). Both work in the same way to predict the presence of the disease allele in question, but the neutral sites are more common and hence generally more useful.

crossing over within genes

Until the mid 1940s, people believed that genes were probably the chromomeres, the tiny bumps along the chromosome that make it look like a string of beads, and that crossing over took place only *between* genes. But some experiments with *Drosophila* showed that crossing over might occur not only between genes but within a gene. Suppose there are two distinct mutant alleles at some locus; a fly may carry one allele on one chromosome and the other on its homolog, so it is heterozygous for the two alleles. Such flies show the mutant phenotype, because both copies of the gene are mutated. But they produce some rare wild-type offspring, which can only arise through recombination. This means that a gene is not an indivisible bead but a linear stretch of a chromosome; that different alleles of a gene could be made by mutations at many places in this stretch; and that recombination can occur between them. We'll represent two such alleles by *1* and *2* and the normal parts of each gene by + signs, the standard symbol for wild-type, although we only place them opposite the mutated sites for contrast to those sites. Then a heterozygote for both alleles has the form:

$$\frac{\underline{\;1 \quad + \;}}{\underline{\;+ \quad 2\;}}$$

leaving a short space between the two mutated sites where a crossover might – rarely – occur. Such an event would produce one copy of the gene that is entirely wild-type and one copy that bears both mutated sites, a double mutant:

$$\frac{\underline{\;+ \quad + \;}}{\underline{\;1 \quad 2\;}}$$

Rare wild-type individuals must arise from such an event.

By studying such rare events, we can map alleles within a gene. But intragenic crossovers are so rare – perhaps one in 5000–10,000 meioses – that mapping genes in flies means counting a lot of flies. Further, to really dissect a gene we need a system in which we can easily manipulate many individuals.

This general method of gene mapping is a very powerful one that has now been used for many organisms and viruses. When combined with more biochemical techniques that we shall discuss later, it has brought some viruses to the point where we essentially know everything about the structure of their genomes, even if we don't know the function of each gene. Next we shall show how the use of phages has provided powerful insights into gene structure.

phage genetics

Max Delbrück began to work with phages because he could see that they provide a very simple biological system: tiny particles that reproduce and must therefore possess genes that they can pass on to their offspring. The first serious attempts at genetic work with phage were made by Hershey, who showed that different strains of phage T2 can recombine. For this work, of course, he needed genetic differences, and the first phage mutants he found were marked primarily by the forms of their plaques. For instance, one can find *r* (for *rapid lysis*) mutants that form a somewhat larger plaque than wild-type, with a very sharp edge; *turbid* (*tu*) mutants make a plaque that is not as clear as normal; and *minute* (*mi*) mutants make tiny plaques. Whenever a plaque is found that has an apparently useful phenotype, the phage in it can be purified and grown up to make a pure genetic strain with a distinct, inheritable phenotype.

A bacterial cell can be infected by several phages simultaneously. Hershey performed a cross by infecting cells with a mixture of an *r* mutant and a *tu* mutant, using enough phages so almost every cell was infected by several of each type. Most of the progeny phages that emerged from these bacteria were either *r* or *tu*, like their parents, but some were double mutants, *r tu*, and some were wild-type. Thus even the genomes of viruses can interact so that crossovers occur between two DNA molecules, producing recombinants. Hershey used several independent mutants, and by interpreting the frequency of recombination as an indicator of distance, just as in classical genetics, he was able to arrange their sites of mutation in a consistent linkage map. The map has now been enormously refined and expanded.

fine structure of genes

Seymour Benzer explored the fine structure of genes by using a T4 phage system in which he could *select* the rare recombinants that occur within a gene. Benzer focused on a class of *r* mutants called *rII*. They grow and form large plaques on *E. coli* strain B but do not grow at all on *E. coli* strain K. In contrast, the rII^+ wild-type forms plaques on both B and K. Benzer found hundreds of new *rII* mutants that proved to be useful not only in mapping but also in providing an operational definition of a gene, based on mutations in the gene.

In a typical mapping experiment, strain B bacteria are infected with two different *rII* mutants, yielding many phages that are mostly the same as their parents – one mutant or the other – with a few recombinants. The total number of phages produced is determined by plating on strain B. But by also plating the phage on strain K, where only wild-type recombinants will grow, the vast numbers of mutants are passed over, so the much smaller numbers of recombinants can be seen and counted.[1] Benzer showed that recombination generally occurs between different alleles within the *rII* locus, and he could determine the genetic distance between any two mutational sites and construct a linear map of these alleles. A small part of the map looks like this:

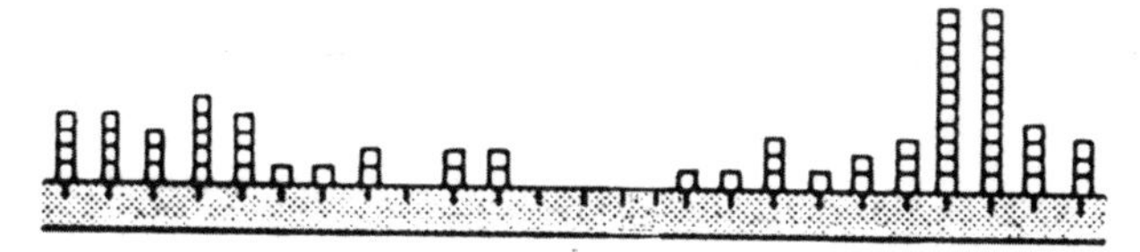

Each square represents an allele separated from other alleles; squares piled up at one point are alleles that could not be separated and therefore represent recurring mutations at the same position. It is clear that with the high resolution of Benzer's system a gene can be dissected into many separate sites, and presumably each site is one nucleotide pair in a DNA molecule.

Benzer's mapping data also make an important point about the structure of a gene. Knowing that genes are made of DNA, everyone assumed that the sequence of bases in the DNA is simply read off in a linear sequence to specify the structure of proteins. But a gene might actually be a little knot or ball of DNA that encodes proteins in some complicated way. Benzer's results say that this isn't so. They say that a gene is a simple, *linear* structure, which is consistent with the simplest hypothesis about how DNA is used.

complementation and the definition of a gene

The mapping experiments showed that the *rII* region consists of many sites where mutations can occur. But there is no real indication of gene *function* here – only of a general gene structure. We don't yet know whether the *rII* region consists of one gene or two or more. A different kind of experiment is needed to identify genes and define their limits, one that has nothing to do with crossing over and mapping, even though it looks superficially like a mapping experiment. The experiment is called a *complementation test*, and it is best explained by setting up a model.

Suppose the *rII* mutations affect two distinct genes that happen to lie next to each other, and both give the same phenotype when mutated. We picture a gene as a region that specifies the structure of one polypeptide, so these two genes must carry the information for two distinct polypeptides; call them A and B. We assume that both genes are needed for normal function in K cells (they seem to be dispensable in strain B), so if some K cells are mixedly infected by two different mutants we should be able to tell whether both proteins are being made. Figure 8.2 shows how different mutations might affect these genes. Suppose the two mutations are both in the *A* gene; since there is no functional *A* protein, the phage cannot grow. Now suppose one mutation affects the *A* gene and the other the *B* gene; now one phage has a functional *B* gene and the other has a functional

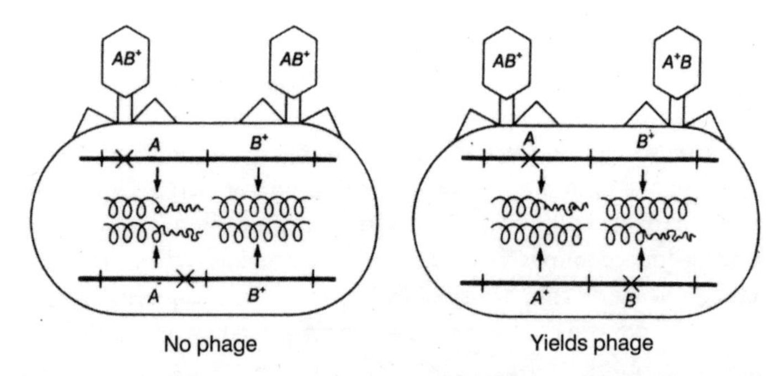

Fig. 8.2. *A complementation test determines whether or not two mutations lie in the same gene. Bacteria are infected simultaneously with phages carrying two distinct mutations, which are either in the same gene (left) or in different genes (right). If the mutations affect the same gene, neither phage has a good copy of that gene, so they cannot multiply. But if the mutations are in different genes, one phage has a good* A *gene, the other a good* B *gene, and the phages complement each other. Notice that crossing over has nothing to do with this test.*

A gene. If a cell were infected simultaneously with these two phages, they could *complement* each other: each one could supply a function missing in the other, and they would both be able to grow. (We need to emphasize again that this test has to do only with the way mutant genes operate; there is no crossing over and no recombination here.)

When Benzer did mixed infections of *rII* mutants two at a time on *E. coli* K, he found exactly the results this model predicts. The mutational sites are spread out along a line, and a boundary in the middle of the region divides them into two groups. None of the mutants that map in the left part of the region will complement one another, and the same goes for those in the right part. But any mutant from the left part will complement any mutant from the right part. This result shows that the *rII* region does consist of two genes. (Although Benzer called the two functional regions *cistrons*, a cistron is really the same as a gene.) Complementation tests like this are now used in all organisms as a standard way to determine whether two mutations are in the same gene or not, and thus to delimit the gene.

what is a gene?

It is now time to revisit the question of just what a gene is. Classically, genes have been defined by three criteria: function, mutation, and recombination. We begin with the conception that a gene is a functional unit, a unit that determines some characteristic; this is still an important part of the gene concept, but we now know that many genes can affect the same characteristic and produce the same phenotype when mutated. A gene is also considered a unit that can mutate. Benzer's experiments provide a picture of a gene as a linear region containing many sites where independent mutations can occur, and we have just shown how genes can be defined experimentally (operationally) by the mutations that occur in them; delimiting a gene operationally by complementation tests is based on the conception of a gene as a unit that encodes the structure of one polypeptide chain, a conception derived from Beadle and Tatum and all the molecular analysis that followed them. Genes have also been defined as units that are separable by recombination; however, it is now clear that a gene is not an indivisible bead on a string and that recombination can occur within genes. This is exactly what we would expect if a gene is simply a stretch of DNA, where any nucleotide pair can be changed by mutation, and where different DNA molecules can be recombined by crossovers.

In the light of contemporary work, especially based on DNA sequencing, we will have to modify our conception of a gene. We will find that in eucaryotes the sequences that encode protein are interrupted by non-coding sequences called introns, which must be removed as a preparation for protein synthesis. Sometimes the coding sequences separated by the introns are combined in different ways to produce different functional proteins from the same DNA sequence; in these situations, if we wish to identify a gene with a single protein product, we will have to say that the one DNA sequence contains several genes. That is one complication. A second complication is that the expression of a gene is controlled by adjacent regions of DNA outside the coding sequences. Mutations in these controlling regions can lead to loss of gene function just as mutations in the coding regions do, so if we are to define a gene on the basis of mutation, we will have to include such control sequences as part of the gene. Finally, the detailed DNA sequences of whole genomes, including now the human genome, are providing the ability to identify genes – at least, likely genes – on the basis of their sequence, rather than on the basis of mutations. Proteins with similar functions, even in vastly different organisms, tend to have much in common; vast databases are now accumulating information about DNA and protein sequences, and computer programs are being developed that can scan a newly determined sequence and identify it as a probable gene whose protein product probably has a certain function. Even if a new sequence does not match anything in the database, we can still infer it to be a gene on the basis of some of some well-known landmarks common to all genes. (We will study these in later chapters.) Based on sequencing, the first analyses of the human genome estimate that it contains about 30,000–50,000 genes, but if one sequence can contain more than one gene, the number of distinct genes may be higher.

The genetic experiments of Benzer and others were critical in developing our picture of gene structure. But it is characteristic of science that advances in a field and new techniques lead to revisions in even the most basic concepts. To get the picture of gene function clear, we must turn in Chapter 9 to more biochemical work that shows precisely how the DNA code is translated into protein structure. Before doing so, we will jump ahead several years to introduce another kind of mapping procedure based on contemporary biochemical manipulation of DNA.

restriction enzymes and palindromes

Bacteria and the phages that attack them are engaged in continual chemical warfare. Bacteria that can resist phage infection are naturally selected because they are more likely to survive; similarly, phages that can overcome the bacterial barriers have an advantage. Bacteria make enzymes called *endonucleases* that attack DNA molecules by cutting some of their phosphodiester bonds. ("Endo-" means they cut somewhere inside a molecule, not at its ends.) These enzymes constitute a *restriction* system that attacks and degrades incoming phage DNA. There are now simple, rapid ways to determine the sequence of a DNA molecule (see Box, pp. 140–42). DNA sequencing shows that each endonuclease is very specific and that it only cuts DNA in a short, specific sequence, most often a *palindromic* sequence. A palindrome is a sentence that has the same letter sequence when read forward and backward, such as Adam's introduction to Eve, "Madam, I'm Adam," or Napoleon's lament, "Able was I ere I saw Elba." A molecular palindrome, which reads the same from either direction, might look like this:

3′-GAATTC-5′
5′-CTTAAG-3′

An enzyme that attacks this particular sequence is made by *E. coli* strain RY13, so if a phage with this sequence in its DNA tries to infect these bacteria, the enzyme will attack that DNA, cut it into fragments, and stop the infection. We now have many restriction enzymes, isolated from different kinds of bacteria, that cut a great variety of palindromic DNA sequences. The source of each enzyme is preserved in a three-letter name abbreviated from the bacterial name, so the one derived from *E. coli* RY13 is named *Eco*RI.

DNA SEQUENCING

DNA fragments may be separated by a process called *electrophoresis* (Fig. B8.1), using a gel material called agarose, which is similar to the agar used to make nutrient media. A mixture of DNA molecules is placed in an agarose strip. When an electric current is passed through the gel, the negatively charged DNA molecules are pulled toward the positive electric pole, the shorter molecules moving faster. The method is so sensitive that it can separate molecules differing in length by a single base. The DNA is visualized by staining the gel with a dye that binds specifically to DNA.

The basic principle of DNA sequencing, invented by Frederick Sanger, depends on two ideas: first, that a DNA strand is synthesized by starting with a short primer and lengthening it in the 5′→3′ direction; second, that a nucleotide can be added to the chain only if the ribose on the end has an oxygen atom on the 3′ position. However, we can make *dideoxy* nucleotides (ddN) that lack this oxygen; they can be incorporated into growing chains, but they terminate the chain because another nucleotide cannot be linked to them. So we start with a single-stranded DNA to be sequenced and add a short, complementary primer that will bind to it, along with DNA polymerase and a mixture of the four nucleotides needed for DNA synthesis. (The mixture contains many molecules of each substance.) The 5′ end of the primer is

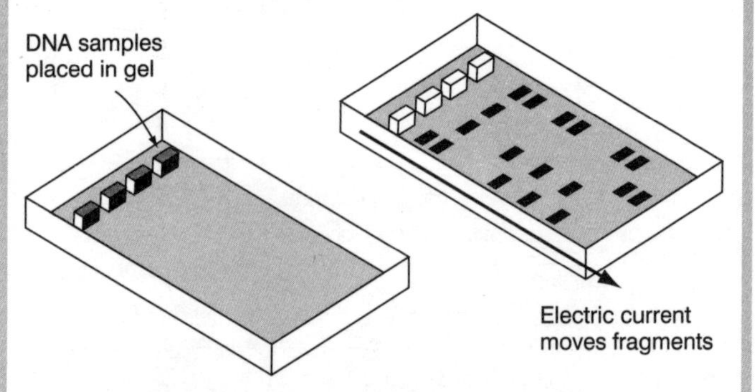

Fig. B8.1. *DNA molecules are easily separated by electrophoresis in an agarose gel. The smallest fragments move fastest.*

generally labeled with ^{32}P, so we can locate the DNA by autoradiography. We then separate this mixture into four tubes. To one, we add a little ddA; to the next, ddC; to the third, ddG; and to the fourth, ddT. DNA synthesis occurs in each tube, extending the primers, but in each tube the molecules terminate randomly wherever a dideoxy nucleotide is incorporated. Thus, if a sequence has ten places where T is to be added, an ordinary dT will generally be incorporated in most places in the tube containing ddT, but enough ddT molecules will be incorporated to create molecules of ten distinct sizes, each ending in T. When the DNA sequences in each tube are separated in a gel, there will be ten distinct bands in the gel from the ddT tube, and similarly for the other tubes. The whole sequence of the DNA can be read directly by starting at the shortest band, identifying its end nucleotide by the tube it came from, and then moving upward band by band (Fig. B8.2).

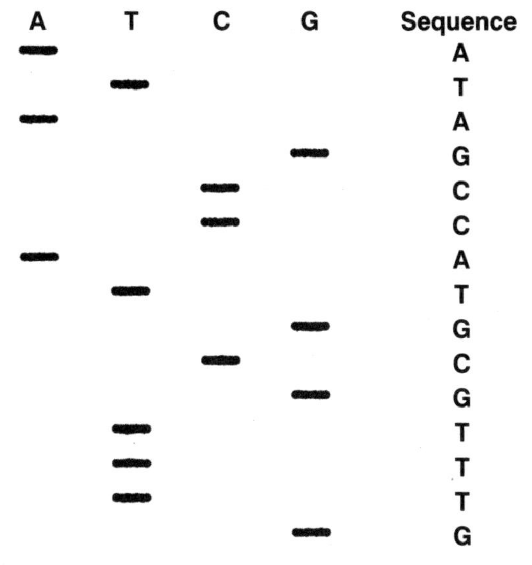

Fig. B8.2 *One method for determining the sequence of a DNA molecule. DNA molecules complementary to a purified single-stranded molecule are synthesized in four tubes, each containing a different dideoxy nucleotide. The resulting fragments are then separated according to size, and one can then read off the sequence from the positions of the bands.*

(continued overleaf)

(continued)

To facilitate handling the enormously long sequences of chromosomes, this method has now been automated. The primers used in the four tubes are labelled with four distinctive dyes, each of which fluoresces with a different color. The gel is then passed through a reading device that irradiates the bands with a laser light and records the color of each band. This information is then fed into a computer, which records the whole sequence. Today in the large sequencing centers, phalanxes of robots perform virtually all the steps in sequencing; culturing cells, DNA extraction, DNA sequencing, and processing sequence data.

How do bacteria avoid destroying their own DNA? They have a parallel *modification* system, using an enzyme that adds a methyl group (CH_3) to the A's in the sequence, thus blocking the action of the restriction endonuclease. Some phages methylate their own DNA so they become resistant to such endonucleases.

restriction mapping

Many types of restriction enzymes are now available. They cut DNA at different sequences and can be used to analyze DNA structure with a different kind of mapping, which is called *restriction mapping.* A restriction map shows the sites of the various nuclease cuts relative to one another, and by using other techniques one can tie this map to a genetic map. The method depends on the technique of electrophoresis explained in the Box (pp. 140–42), which is used to separate DNA fragments and determine their sizes. Restriction mapping is best explained by an example. Many small animal viruses have small, circular DNA genomes. Suppose we isolate some viral DNA that is about four kilobases (kb) long and cut it with the enzyme *Eco*RI. When we run this DNA on a gel, we find four fragments: 0.4, 0.8, 1.3, and 1.5 kb. This means that there are four *Eco* sites on the genome, and they could be arranged in several ways.

We again cut some of the viral DNA with *Eco*RI, but using a short digestion time so some of the DNA is incompletely cut. This gives the same four fragments, along with some longer fragments: 1.7, 1.9, 2.1, and 2.3 kb, among others. A little experimenting with possible maps shows that we should expect these fragments if the cuts are arranged like this:

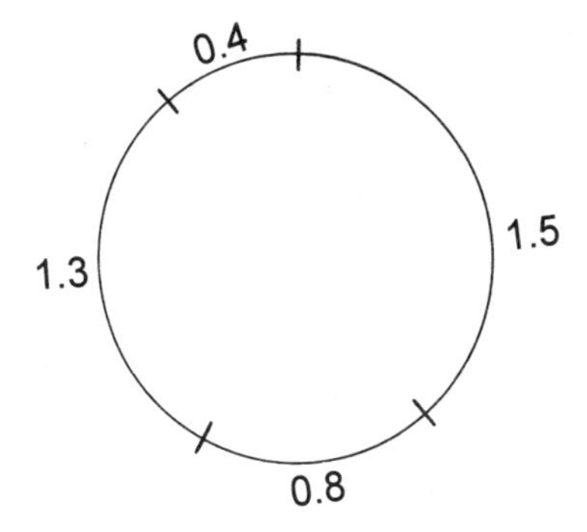

We can confirm this by isolating some of the larger fragments, cutting them further with *Eco*RI, and seeing if they yield the expected smaller fragments.

Next, we cut some more of the same viral DNA with the enzyme *Sal*I and find three fragments: 0.95, 1.25, and 1.8 kb. So there are three *Sal*I sites. Then we isolate each *Sal*I fragment and cut it with *Eco*RI:

0.95 kb fragment ⟶ 0.1 kb + 0.85 kb
1.25 kb fragment ⟶ 0.2 kb + 0.4 kb + 0.65 kb
1.8 kb fragment ⟶ 0.7 kb + 1.1 kb

A little experimentation with these numbers shows that the *Sal*I cuts can be arranged relative to the *Eco*RI cuts like this:

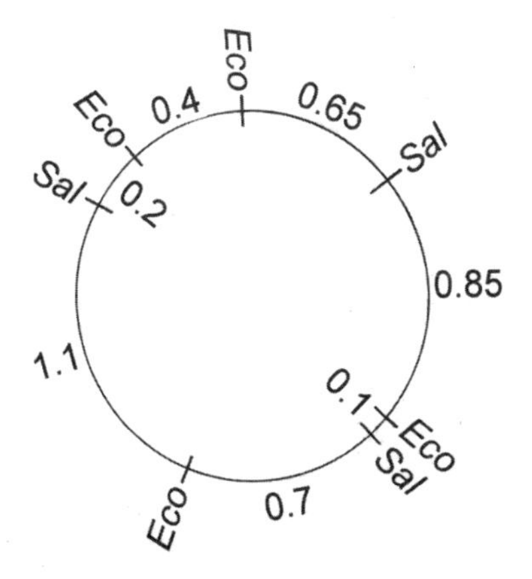

If there were not enough information to make a reasonable judgement, we could get more by partial digestion with one enzyme and then cutting with the other, or by cutting first with *Eco*RI and then with *Sal*I. We can then introduce a third enzyme and map its cuts relative to the other two by an extension of the same methods. In fact, this is a particularly simple example; most genomes, even of small viruses, would yield considerably more fragments and would demand more complex analytical schemes. But the principles of analysis are always the same.

An exciting application of this technique is the diagnosis of hereditary disorders from the DNA of fetal cells recovered by amniocentesis. We find that the enzyme *Hpa*I produces different fragments from the DNA of the normal and sickle cell anemia alleles of the hemoglobin beta gene; eighty-seven percent of sickle cell anemia samples yield longer fragments than normal. This means that the HbS mutation abolishes an *Hpa*I cleavage site. Using this method, it is possible to detect eighty-seven percent of sickle cell anemia babies *in utero*.

Restriction enzymes are useful in another way too. All species have minor variations in DNA sequences among individuals, and in some places these variations produce or eliminate specific restriction sites. This is usually neutral variation that has no effect on phenotype – not like the difference in the middle of the hemoglobin gene described above. These neutral variable sites are useful in mapping, because they create alternative forms of a chromosome, with or without a potential cut site for a restriction enzyme:

Cut ______________________________ Cut

or

Cut __________ Cut __________________ Cut

Molecular techniques for probing this region will reveal this difference, because when these DNA sequences are cut with a restriction enzyme, one will produce one large fragment and the other will produce two smaller fragments whose total size is that of the large one. Thus, there are two *morphs* in the population – some individuals have the cut site and some do not – so the variable site produces a *restriction fragment length polymorphism* (*RFLP*, pronounced "rif-lip"). RFLPs are valuable in mapping because each of them provides a neutral heterozygous marker that can be used to map genes in the vicinity, essentially like mapping markers that produce different phenotypes. Furthermore, a RFLP very close to a disease allele can be used as a diagnostic for the presence of the allele, much like the restriction site that is in the middle of the hemoglobin gene. Genomes contain other types of neutral variable sequences, but RFLPs were the first to be discovered. They can all be used as markers for mapping and for showing the presence of recessive disease alleles.

chapter nine

deciphering the code of life

As soon as Watson and Crick proposed their model, people realized that the linear sequence of bases in DNA could form a series of code words, or *codons*, that would specify the corresponding linear sequences of amino acids in proteins. Furthermore, as Crick pointed out, since DNA and protein are both linear molecules, the two sequences should be *colinear*. This means that the first codon in a gene should encode the first amino acid, the next codon should encode the second amino acid, and so on. This principle still left room for some armchair logic and for many ingenious proposals about the way a sequence of A, G, C, and T in various orders might be used as a code.

Proteins contain twenty different amino acids. Suppose two bases in sequence, such as AA or CT, code for one amino acid. Since there are four ways to choose the first base and four ways to choose the second, there are $4 \times 4 = 16$ combinations. Sixteen isn't enough. Suppose, then, that a triplet, like ATT or GCT, is used as a codon. Now there are $4 \times 4 \times 4 = 64$ combinations. Sixty-four is too many, but at least a triplet code would work. Perhaps forty-four of the triplets are meaningless or – more likely – perhaps the code is *degenerate*. That's not a moral indictment; it means that several triplets could be codons for the same amino acid. Another possible coding mechanism, which was seriously considered for a while, is that some of the bases might make a code while others make "commas" that separate the codons from one another.

How do we know which of these alternatives is right? Crick and his associates attacked the problem directly with a brilliant series of experiments using the *rII* mutants of phage T4, because it is so easy to determine whether any phage has the mutant or wild-type phenotype; remember that *rII* mutants do not grow on strain K. This work depends on the mutagenic effects of the dye *proflavin*, whose molecules tend to slip between adjacent bases in the DNA double helix. Then when this DNA is replicating and recombining, it produces molecules that have insertions of one or a few extra nucleotides or deletions where one or a few are missing. To simplify the situation, we'll assume that every proflavin-induced mutation is an insertion or a deletion of just one nucleotide.

Although you don't realize it, when you read you use a *reading frame*. It is like a little box that you move along step by step to mark off the words, which are conveniently marked by spaces. Your frame has to expand and contract because the words are different lengths, but if you knew that every word were, say, three letters long, you could use a frame of that length and read the sentence

THE FAT TAN CAT ATE THE BIG RAT

even if it had no spaces. But suppose an extra letter were inserted; this would shift your reading frame so you would read:

THE FAT TAN CAT JAT ETH EBI GRA T . .

and so on. From the point of the insertion, the words are gibberish, but some of the meaning could be rescued with a compensating deletion that would shift the reading frame back:

THE FAT TAN CAT JAT ETH BIG RAT

We can at least see that the story involves a fat cat and a big rat, in spite of the gibberish between the insertion and the deletion.

This model is an analogy for a DNA molecule carrying a string of codons that are three bases long, where a *frameshift mutation* can move the reading frame in one direction or the other – which we can arbitrarily designate R and L. This is what Crick and his friends had in mind as they did their experiments. They began with one proflavin-induced mutant, called FC0, which they arbitrarily called a frameshift in the R direction. They could then select compensating L mutants. FC0, being an *rII* mutant, won't grow on strain K; but if we mutagenize some of these phages with proflavin and plate them on K, some phages do make plaques because they have a compensating L mutation. This second mutation is a *suppressor* of FC0: a suppressor is a second mutation that nullifies the effect of some other mutation. And so the Crick group collected a number of these suppressors (numbered FC1, FC2, and so on), and by crossing these double mutants (carrying R and L mutations close together) with wild-type phages, they isolated the L mutants. Since FC1, FC2, and the others are all L mutants, any suppressors of them would have to be R mutants. By going back and forth in this way, the team collected several R mutants and L mutants.

The very ability to collect mutants of this kind fits the model, but there are two critical tests. First, any L mutant should be able to suppress any R mutant – at least, if their sites are close to each other – so the Crick group created many such pairwise combinations. Indeed, they found that, with a few exceptions, double mutants containing one R and one L mutation appear to be wild-type. But the most telling experiment was to put together three R's or three L's. You can easily prove to yourself that if the code is really triplet, three shifts in the same direction will undo one another and restore the correct reading frame again. And this is what they found: a triple-R or a triple-L mutant generally has the wild-type phenotype if the mutations are fairly close together. This was a stunning demonstration that the code is triplet and that it is read by a reading frame mechanism without commas.

Furthermore, this system can only work if the code is degenerate. Some wrong triplets appear between an R and an L mutation or in the space covered by three R mutations together. But the phenotype is generally not mutant because all sixty-four triplets, or at least almost all, must actually code for some amino acid. Even if a double or triple mutant has a stretch of the wrong amino acids, the structure must be good enough to make a functional protein. If only twenty triplets were codons and the other forty-four were nonsense, any arbitrary mutation would probably create one of those forty-four, and the protein synthesis apparatus would stop whenever it encountered a nonsense triplet. Since this does not usually happen, it argues for degeneracy of the code. In fact, as we shall see, only three of the sixty-four triplets are nonsense.

how are proteins made?

So the information specifying the amino acid sequences of proteins is stored in DNA as a series of triplet codons. But how is the sequence of bases in DNA converted into the actual product? Having a blueprint or set of instructions is important, but converting them into a finished product can require a complex operation.

The very structure of eucaryotic cells dictates one limitation. The DNA is located in chromosomes in the nucleus, whereas proteins are manufactured in the cytoplasm on particles called *ribosomes*, which mostly cover the surfaces of endoplasmic reticulum membranes (Figure 9.1). If DNA carries the blueprint for making proteins, but the protein-synthesizing apparatus is in the cytoplasm, how do the instructions get to the apparatus? It's like getting a vital formula from a rare book that you are not allowed to take outside the library. The solution, of course, is to photocopy the page and take the copy to the workshop. That is what a cell does. The copy is RNA.

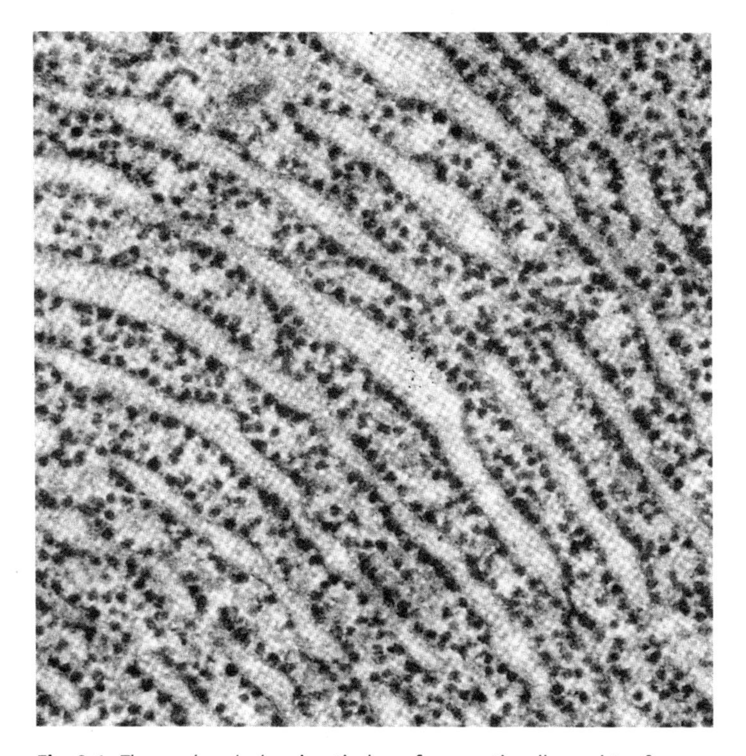

Fig. 9.1. *The rough endoplasmic reticulum of eucaryotic cells consists of membranes, generally parallel to one another, covered with minute particles called ribosomes, which are the factories where protein is synthesized.*

rna molecules: the tools for protein synthesis

During the 1940s, even before people understood the structure of nucleic acids well, there was evidence that protein synthesis is always accompanied by ribonucleic acid (RNA) synthesis. As we showed in Chapter 7, RNA differs from DNA in three ways: its sugar is ribose rather than deoxyribose; it is generally single-stranded, rather than double-stranded; and the thymine of DNA is replaced by uracil in RNA. However, uracil is almost the same as thymine, and it can bond to adenine to make an A–U base pair just like the A–T base pair of DNA. A single-stranded RNA molecule can fold back on itself to make double-stranded hairpin regions held together by A–U and G–C base pairs.

Why would organisms evolve two types of nucleic acid that are distinct yet so similar? One answer is that both RNA and DNA probably developed out of the "primeval soup" when early life forms were evolving, so they were both naturally there; it was then a question of whether they were useful in primitive cells or not. There is good evidence that the first primitive genomes were composed of RNA, not DNA. Eventually this genomic RNA world gave way to a world of DNA genomes, for reasons that are not clear. Through a selective process, the two molecules became specialized in function – DNA as the genome, RNA molecules as the tools for protein synthesis.

Definitive evidence that RNA carries information from DNA to the cytoplasm came from later *pulse-chase* experiments using radioactive tracers. This kind of experiment can follow the fate of a substance just as you could follow the flow of water in a river by throwing in a dye marker to make a colored spot and then following the spot to see where it goes. We expose some eucaryotic cells briefly to uridine labeled with tritium, the radioactive isotope of hydrogen (^{3}H-uridine). This uridine is specifically incorporated as a label into whatever RNA the cells make during this short time. Because of the short exposure time, this is called a *pulse label.* To end the labeling period, we soon *chase* the RNA by diluting the ^{3}H-uridine with massive amounts of unlabeled uridine. We can then trace the fate of the labeled RNA by removing some cells at various times and preparing them for autoradiography (see p. 125), so the dark grains made by the radioactive emissions from the tritium show where the RNA is. Initially, all the label is found in the nucleus. Then while the amount in the nucleus decreases, more label appears in the cytoplasm. This tells us that RNA is made in the nucleus and moves into the cytoplasm. Furthermore, the longer the time between giving the pulse and making the autoradiogram, the fewer grains appear. This tells us that the RNA made during the pulse must break down over time. We say that the RNA *turns over*: it is made, used briefly for some function, and then broken down, rather than remaining as a stable component of the cell. (The idea of something "turning over" isn't confined to biology. Store owners want their inventories to turn over quickly, as they order new goods from their suppliers to replace what customers buy off the shelves. The residents of hospitals and hotels turn over at various rates, too.)

Another bit of evidence came from the work of Elliot Volkin and Lawrence Astrachan, who examined the nucleic acids made in a bacterial cell after phage infection. They identified a new kind of RNA whose base composition is strikingly similar to that of the phage DNA but different from the bacterial DNA. This RNA also turned over rapidly.

A nucleic acid may be characterized by its *base ratio* (A + T)/(G + C). The base ratio of DNA from various organisms varies enormously, and this number has been used to identify the most closely related species and to classify them, on the assumption that related species should have very similar DNA. In contrast, the RNA of these organisms has a pretty constant base ratio. The bulk of this RNA is *ribosomal RNA* (*rRNA*), large molecules that make much of the structure of ribosomes. A ribosome is a large particle, made of two unequal pieces. Each piece consists of 30–40 different proteins, some of which are enzymes that help to link amino acids into new protein chains; they are attached to a core of long rRNA molecules, so the ribosome is rather like an irregular blackberry or raspberry. Most of the other RNA consists of smaller molecules known as *transfer RNA* (*tRNA*), whose function in protein synthesis we'll explain shortly. However, these RNAs do not turn over. Once made, they are quite stable. So the stuff found by Astrachan and Volkin and the material labeled by a pulse in eucaryotic cells appears to be a different kind of RNA. Sydney Brenner, François Jacob, and Matthew Meselson used another kind of labeling experiment to show that the

RNA made after phage infection attaches to the ribosomes for a short time and is then broken down. They argued that this RNA must be carrying a message from the phage DNA to the ribosomes and called it *messenger RNA* (*mRNA*). The ribosomes are merely factories that make protein; they attach temporarily to messenger RNAs, which program them to make specific proteins.

rna transcription

It is now clear that RNA is made on DNA by the same complementary base pairing that produces a new DNA double helix from a single strand (Figure 9.2). This process is called *transcription*, and it is carried out by a complex enzyme called *RNA polymerase*. Near each gene is a region called a *promoter*, which has a sequence of bases that the RNA polymerase binds to – that is, which the polymerase *recognizes*. The polymerase then opens the double helix of DNA slightly and moves along the gene, nucleotide by nucleotide, synthesizing an RNA molecule with a base sequence precisely complementary to that of one strand of the DNA (the *template strand*), using the rule that U pairs with A and C pairs with G. The resulting molecule is called a *transcript*. This transcript has a base sequence identical to that of the complementary DNA strand – the

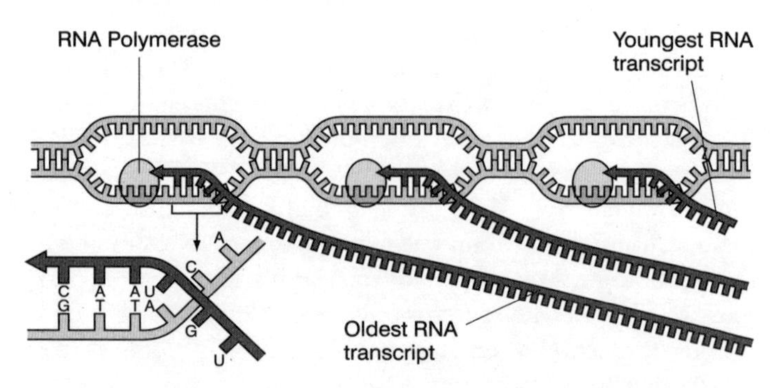

Fig. 9.2. *In the process of transcription, an RNA strand is synthesized on DNA by copying one DNA strand into a complementary sequence. The process is analogous to making a new strand of DNA during replication, but (a) only one strand of DNA is copied, (b) the product is RNA, not DNA, and (c) the process is carried out by the enzyme RNA polymerase.*

coding strand – except for having U instead of T. The relatively simple process of transcription we have described here occurs in bacteria; the formation of mRNA in eucaryotes entails some complications, discussed in Chapter 11.

With sophisticated techniques of electron microscopy, DNA can be captured in the act of transcription (Figure 9.3). Transcription can also be carried out *in vitro* – in a test tube – using purified DNA plus RNA polymerase and the four kinds of RNA nucleotides.

A critical demonstration of the base sequence complementarity between DNA and RNA comes from techniques in which nucleic acids are hybridized with one another. In 1960, Paul Doty and Julius Marmur found that at high temperatures the hydrogen bonds that hold a DNA duplex together are disrupted, so the DNA *denatures* and the strands of the duplex separate. If the solution of denatured DNA is allowed to cool slowly, the single strands have time to find other strands with complementary base sequences, and these come into register so the strands reanneal to form stable duplexes again. Furthermore, a DNA strand can anneal with an RNA strand to form

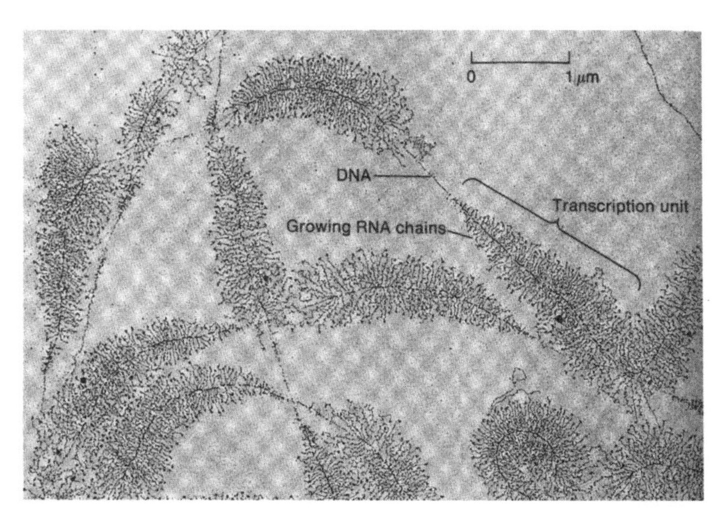

Fig. 9.3. *An electron micrograph of DNA being transcribed into RNA. The DNA is the thin strand running through the center of each region with the shape of a feather. The strands leading off the DNA are RNA molecules; the longest ones are nearing completion and the shortest ones have just been started. These are ribosomal RNA molecules that are incorporated into the structure of ribosomes.*

a hybrid molecule. Reannealing is incredibly specific and demands exact complementarity between two strands. Since an RNA molecule is perfectly complementary to the DNA strand it was made on, the two should anneal in the right conditions. The standard technique is to denature DNA and bind the resulting single strands to thin nitrocellulose filters. When these filters are incubated with labeled RNA, any RNA transcripts with complementary sequences will bind to the DNA and are easily detected by measuring radioactivity on the filter; RNA molecules that do not have complementary sequences do not bind and are easily rinsed away.

A DNA molecule has two strands. We showed above that only one strand – the template strand – is transcribed by each RNA polymerase, but are both strands transcribed at different times to make two RNA molecules? Julius Marmur provided one answer to this question with a phage called SP8; its DNA is one duplex molecule whose strands are very different in density (the more G a strand has, the denser it is), so they can be separated in a CsCl gradient. He found that RNA made after SP8 infection hybridized with only one of these strands. This shows that only one strand of the DNA acts as a template for RNA synthesis. In other viruses and in cells generally, both strands are used in different regions, but usually only one strand is transcribed in each region. A bit of reflection shows that this must generally be true, because RNA molecules made from the complementary strands of one gene would encode very different proteins, and only one of them could be a functional product of that gene.

We now have to further broaden our concept of a gene, for a functioning cell has two kinds. Most genes carry the information for a particular kind of protein; transcripts from these genes are the messenger RNAs, which direct protein synthesis. The sequences of the stable RNAs – the ribosomal and transfer RNAs – also have to be specified by DNA, so the genome contains genes for the various tRNA and rRNA types. The products of these genes are not proteins but, rather, RNA transcripts that are stable cell components, part of the protein-synthesizing apparatus, as we will now show.

the translation process

The transfer of information from DNA to RNA is transcription; the transfer of that information from the mRNA to protein is *translation.* A messenger RNA programs each ribosome for a short time, and generally the mRNA is then broken down. The ribosome itself establishes the proper reading frame by pinching the mRNA between its two subunits and moving it along like a ratchet, exposing three bases at a time, from the 5′ to the 3′ end. In fact, several ribosomes can bind to a messenger one after another, so they are all synthesizing the same protein one after the other. Several ribosomes attached to an mRNA form a *polyribosome.*

The mRNA contains a series of codons, copied from the DNA; but an amino acid does not have the chemical structure needed to bind specifically to a codon – to recognize the codon. This is where the transfer RNAs come in (Figure 9.4). Cells have at least one kind of

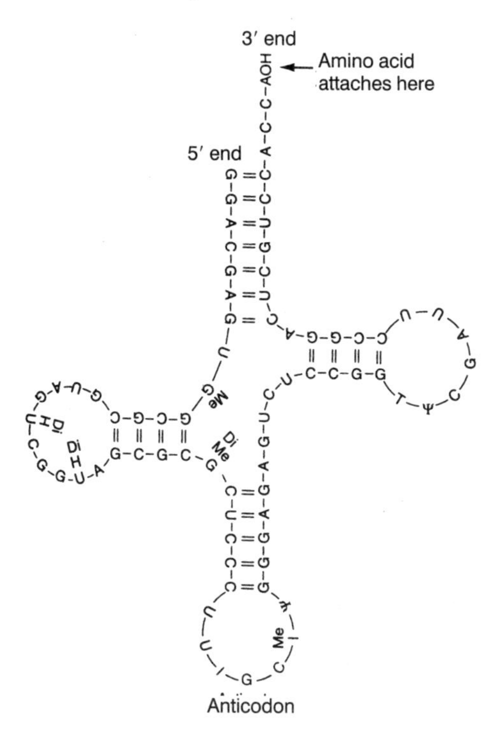

Fig. 9.4. *General structure of a transfer RNA molecule. Notice the extensive internal bonding that forms G–C and A–U pairs. The amino acid is carried at one end, and the opposite end consists of a loop carrying the anticodon that recognizes the codon on the messenger RNA. Some of the bases are modified by the addition of small chemical groups, such as hydroxyl and methyl groups.*

tRNA for each of the twenty kinds of amino acid. The cell also contains twenty different kinds of enzymes (aminoacyl RNA synthetases), one for each amino acid. Each enzyme recognizes a certain tRNA along with its corresponding amino acid and brings them together. It attaches the proper amino acid to each tRNA to make an *aminoacyl tRNA*. Three bases at one spot on the tRNA form an *anticodon* that is complementary to the mRNA codon, so each aminoacyl tRNA can bind to the mRNA at the right place.

Translation (Figure 9.5) begins when an mRNA attaches to a ribosome. Each codon of the mRNA is recognized, in its turn, by a tRNA carrying the proper amino acid, so the amino acids are polymerized into a protein chain in the sequence specified by the sequence of codons. Notice that two tRNAs bind at once; then the amino acid of the first one is bonded to the amino acid on the second, releasing the first tRNA. The ribosome then shifts the mRNA along by one codon, simultaneously moving the second tRNA so a third one can occupy its place. The process is repeated, so the protein grows by one amino acid at a time and is always attached to the last tRNA. (This link is eventually broken when the protein is completed.) This very pretty picture of translation is quite accurate, although numerous details remain to be explained.

the complexity of eucaryotic genes

As investigators began to locate the genes for various proteins in eucaryotic cells, some of them started to get evidence that the relationship between genes and proteins in eucaryotes is more complicated than in procaryotes. Firm evidence for the relationship emerged in 1977 in the laboratories of Philip Sharp and Pierre Chambon. They and their coworkers hybridized the mRNAs for specific genes with the DNAs from which those messengers were transcribed. In bacteria, the mRNA sequence is identical to that of the coding strand (except for U in the place of T), so such hybrid structures look very simple. But when these structures for eucaryotic genes were examined by electron microscopy, they showed a series of loops. This means that the mRNA and the DNA do not have exactly the same sequence, and the loops are regions where they cannot hybridize. When the mRNA sequences were compared with the DNA sequences, it

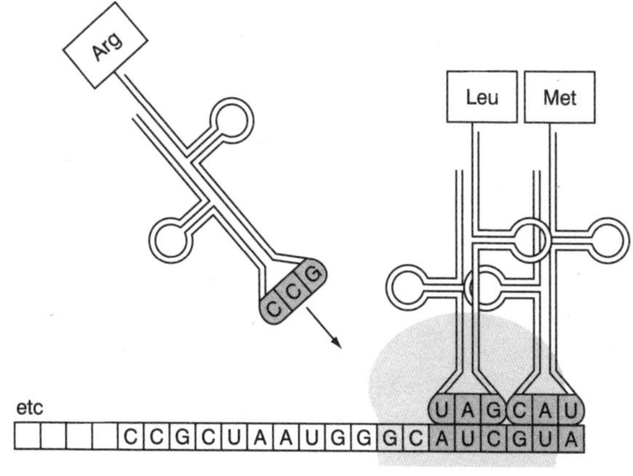

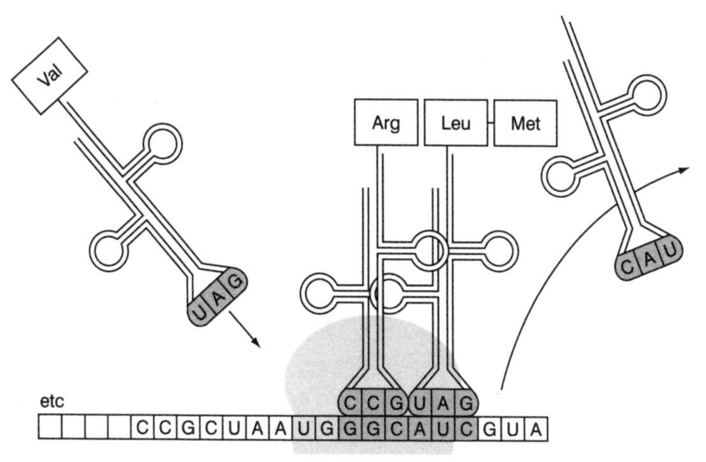

Fig. 9.5. *An outline of protein synthesis. The messenger attaches to the ribosome so that its first two codons are in binding sites for aminoacyl tRNA molecules. Two aminoacyl tRNAs – those that are called for by the codons – occupy these sites. The first amino acid is then connected to the second by a peptide linkage, leaving the dipeptide attached to the second tRNA. This complex moves along by movement of the messenger on the ribosome, so now a third aminoacyl tRNA can bind. A peptide linkage is made between the second and third amino acids, and the process is repeated. This goes on, generally for hundreds of amino acids, until a stop signal appears on the messenger and the finished protein is released.*

became apparent that the coding sequences of the gene are interrupted in several places by nonsense sequences, series of nucleotides that are never translated into protein. This became a regular pattern, and it is now known to be typical of eucaryotic genes. The coding sequences in the gene are called *exons* ("expressed" sequences) and the interrupting nonsense sequences are *introns* ("intervening" sequences). Some genes are interrupted by several introns, and it is common to find genes with much more intron DNA than exon DNA.

In general, eucaryotic genes are transcribed into large RNA molecules containing both the intron and exon sequences. Then specialized enzyme complexes (splicosomes) precisely cut out all the intron RNA from the transcript, joining the exons into a single mRNA that encodes only the protein. This RNA is translated as usual.

The reason nature has retained this structure is still somewhat elusive, but it can be rationalized by reference to both evolution and development. From the evolutionary viewpoint, this structure is valuable because it allows for greater genetic experimentation, leading to new genes. Crossing over can occur within introns, where mistakes in sequence make no difference, and recombination can produce novel combinations of exons and thus novel proteins. It commonly happens that each exon encodes one *domain* of a protein, a domain being a distinctive part of the protein that has a specific function. Thus, putting together new exons creates proteins with new combinations of domains and thus new potential functions. Such changes in genetic structure are the source of all novelty in evolution.

From the developmental viewpoint, the intron–exon structure is valuable because it allows a single nucleotide sequence to encode more than one protein. Several cases are now known where introns are spliced out in alternative ways in different tissues, producing two or more distinct proteins with different functions. Thus, the structure permits novelty with minimal information.

Eucaryotic chromosomes contain not only extensive intron DNA but also various kinds of repetitive DNA that does not encode either proteins or stable RNA molecules. For instance, about ten percent of mouse DNA is *highly repetitive* and consists of short sequences, less then ten nucleotide pairs long, repeated millions of times. Another twenty percent of mouse DNA is *moderately repetitive* and consists of sequences of a few hundred nucleotide pairs repeated about a thousand times. Thus, much of a typical eucaryotic chromosome consists of DNA that can be altered considerably by mutation and recombination without any obvious effect. Repetitive DNA in the human genome is discussed in Chapter 12.

cracking the code

By 1962, the work of Crick and his colleagues, discussed above, had established that the genetic code is triplet. The obvious problem was to crack the code and identify each triplet with the proper amino acid. As is so often the case, the breakthrough came from an accident; the whole code was then worked out in just a few years – surely one of the most exciting achievements of molecular biology. In 1961, Marshall Nirenberg and Philip Leder were developing an *in vitro* system for synthesizing proteins, a mixture of ribosomes, energy sources, activating enzymes, tRNA, and so on. For one control – a mixture in which no protein synthesis was expected – they added some artificially made RNA composed only of uracil, a polymer with the sequence U–U–U–U–U–, called poly-U. Contrary to expectations, this poly-U acted like a messenger RNA and stimulated protein synthesis. The system with poly-U in it would only synthesize a polymer of the amino acid phenylalanine, which meant that UUU must be the code word for phenylalanine.

This discovery was followed by a race between Nirenberg's and Severo Ochoa's laboratories, using synthetic RNA to find each code word. Since the enzyme that makes such synthetic molecules creates random sequences, the first studies had to depend on a kind of statistical analysis of the resulting peptides. The real breakthrough came when Nirenberg and J. Heinrich Matthaei tried making mini-messengers, only three nucleotides long, with known sequences. They found that in an *in vitro* system each of these triplets would attach to a ribosome and only a single kind of tRNA would recognize it. This made it relatively easy to identify the amino acid coded by each triplet. They found that both UUU and UUC (reading each triplet in the $5' \rightarrow 3'$ direction), for instance, would bind phenylalanine tRNA, that GUU binds valine tRNA, UUG binds leucine tRNA and UGU binds cysteine tRNA. Eventually, this work, with confirmation from other sources, led to the complete genetic dictionary given in Table 9.1.

Table 9.1. *The genetic code*

First base (5′ end)	Second base: U	C	A	G
U	UUU, UUC Phe UUA, UUG Leu	UCU, UCC, UCA, UCG Ser	UAU, UAC Tyr UAA Stop UAG Stop	UGU, UGC Cys UGA Stop UGG Trp
C	CUU, CUC, CUA, CUG Leu	CCU, CCC, CCA, CCG Pro	CAU, CAC His CAA, CAG Gln	CGU, CGC, CGA, CGG Arg
A	AUU, AUC, AUG Ile AUA Met	ACU, ACC, ACA, ACG Thr	AAU, AAC Asn AAA, AAG Lys	AGA, AGG Arg AGU, AGC Ser
G	GUU, GUC, GUA, GUG Val	GCU, GCC, GCA, GCG Ala	GAU, GAC Asp GAA, GAG Glu	GGU, GGC, GGA, GGG Gly

Each of the sixty-four triplets either stands for one of the amino acids (represented by its three-letter abbreviation) or signals the end of the polypeptide chain.

Several points stand out in Table 9.1. As Crick predicted, there is considerable degeneracy, but the number of codons specifying an amino acid ranges from one (methionine, tryptophan) to six (leucine, serine, arginine). Furthermore, the degeneracy is very regular. In general, the first two bases (reading from 5′ to 3′) have most or all of the meaning; in eight cases, once these two are given it makes no difference what the third base is. In twelve cases, it only matters whether this base is a purine (A, G) or a pyrimidine (U, C).

The triplet AUG, coding for methionine, is almost always used at the beginning of a gene to signal a special tRNA carrying a methionine with its amino group blocked (*N*-formyl methionine). Methionine at other places in a protein is carried by a different tRNA.

colinearity of genes and proteins

The hypothesis that a gene is colinear with its protein product could be confirmed by showing that the sequence of mutations in a gene is the same as the sequence of amino acid changes that those mutations produce. This requires a gene that can be mapped and whose protein product has been sequenced. Charles Yanofsky found the right system among the genes for enzymes that synthesize tryptophan (the *trp* genes) in *E. coli*. Yanofsky and his colleagues used the *trpA* gene that encodes the A protein, part of the enzyme tryptophan synthetase. They determined the entire sequence of 267 amino acids in the wild-type A protein and in the proteins made by ten mutants in the *trpA* locus; each mutation produced only one amino acid substitution in the entire protein. They also mapped the mutational sites of the ten mutants and showed that sequences of mutational sites and amino acid changes in the protein are colinear (Figure 9.6). The genetic distance between two mutational sites is proportional to the distance between the amino acids they affect, so genetic distances based on recombination reflect physical distances within a gene. Note also that, in two cases, two alleles that are separable by recombination affect the same amino acid. The normal glycine residue at position 210 is changed to arginine in mutant *A23* and to glutamic acid in mutant *A46*. The genetic code accounts for this if the glycine codon is GGX (X is either A or G), and it changes to CGX in *A23* and to GAX in *A46*. *A78* and *A58* also change the same codon

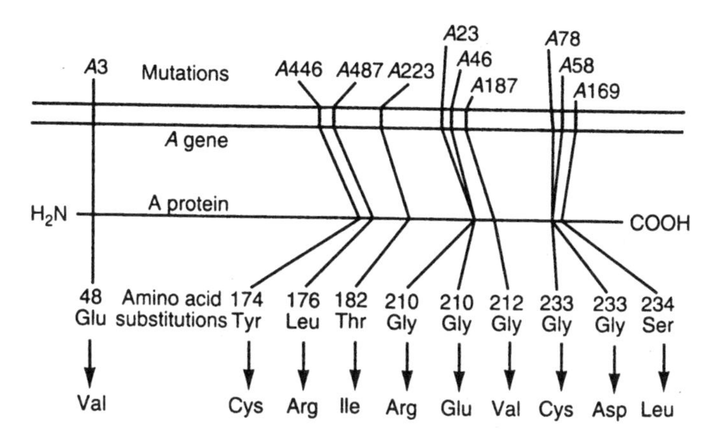

Fig. 9.6. *Colinearity of a gene and its protein product. The sequence of mutational sites in the DNA is identical to the sequence of amino acid changes in the protein. In two cases, two neighboring mutations produce different substitutions for the same amino acid.*

in different ways. Each gene thus consists of a linear series of codons that dictate amino acid sequence, and a coding element for one amino acid is itself subdivisible by recombination.

This elegant genetic experiment was confirmed by another beautiful experiment performed by George Streisinger and his coworkers. They obtained mutants of a phage T4 enzyme called lysozyme, and also isolated the protein and determined its amino acid sequence. After inducing a mutation in the gene with proflavin, they selected a proflavin-induced suppressor. The double-mutant protein turned out to have the wild-type sequence except for a short stretch of amino acids corresponding to the gibberish between the two mutant sites. This was a neat extension of Crick's work.

These genetic experiments were essential in developing our picture of how genes work. Independent of the biochemical work, they confirmed that work and also demonstrated in other ways how the DNA code is translated into protein structure.

stop codons

Three of the sixty-four triplets are stop signals that do not code for amino acids but mark the end of a protein. Their existence was shown by the discovery, in some bacterial and phage systems, of *nonsense mutants* in which a mutation has produced one of these stop triplets somewhere within a gene. Normally, these triplets should occur only at the end of a gene. But if a stop codon is created by mutation in the middle of a gene, protein synthesis stops short. One series of nonsense mutants, the *amber* mutants, is defined by having a defect that can be *suppressed* by a second mutation that affects the protein-synthesizing apparatus itself; the suppressor is a change in one of the transfer RNAs so it can recognize the nonsense codon and insert an amino acid, thus changing nonsense into sense. Sequencing of normal proteins has shown that the *amber* triplet can come from only a few codons such as UGG, UAC, and CAG, from which it is inferred that the *amber* triplet is UAG. This pattern of mutation thus confirms the codons assigned through biochemical work.

universality of the code

The codon assignments are based on pioneering studies using the *E. coli* protein-synthesizing apparatus. Are the triplets thus deciphered valid for humans? If they weren't, mutations observed in human proteins would not be consistent with single base changes in the *E. coli* derived codons. But the sequences of many genes and their corresponding proteins have been determined in a variety of organisms, and the code is the same in every case. (A few rare exceptions have been found. Also, mitochondria and chloroplasts contain their own DNAs that specify some of their components, and they use a few variant codons.) In fact, we can mix and match components; mRNAs from *Drosophila*, for example, can be used with the protein-synthesizing apparatus of plants to get very high-fidelity translation into *Drosophila* proteins. So the code and protein-synthesizing machinery appear to be universal. This universality was one of the first experimental demonstrations that the life forms on this planet are descended from a common ancestor. The work of the great Victorian evolutionists Charles Darwin and Alfred Russell Wallace, and their academic descendents in the early half of the twentieth century, showed many types of evolutionary relatedness between specific organisms. However, few could have predicted the discovery from the early years of molecular genetics that *all* living organisms are derived from a common ancestor. So we humans are kin not only to the other primates, and to animals generally, but to plants, fungi, and bacteria. This realization has vast implications, not only for biology but for all of humanity.

Thus the stage has been set for the exciting molecular engineering that promises to change our relationship with nature in the profoundest manner. How that might be done and with what consequences will form the subject of Chapter 12. First, in Chapter 10, we must explore the genetic systems of bacteria, which are used for manipulating other genomes.

chapter ten

heredity in the bacterial world

Classical geneticists could hardly have dreamt of the new vistas of understanding and opportunity that would be opened up by experimentation with bacteria and bacterial viruses. In this chapter we explore some of the results of this work and the emerging principles that have allowed us to understand how the genetic apparatus operates, not only in bacteria, but in organisms generally. We also encounter some of the new opportunities and challenges for humanity as a whole that have have stemmed from this work. The abilities to clone, sequence, and manipulate DNA, and to create genetically modified organisms, have all emerged directly from pioneering advances made in bacterial and viral genetics.

mutant bacteria

Different species of bacteria can be identified by phenotypes such as the shapes, colors, and minute details of their colonies. But genetic work with bacteria has progressed by picking one species (most prominently, *E. coli*) and finding mutants that differ from the norm in subtle ways. Some of the most useful mutants have a *conditionally* expressed phenotype, meaning that under so-called *permissive* conditions they appear to be wild-type but under other, *restrictive* conditions the same cells show some other phenotype. For instance, *temperature-sensitive* (*ts*) mutants grow normally at relatively low temperatures (say, 28°C) but cannot grow at 42°C. *Cold-sensitive* (*cs*) mutants are just the opposite. Thus by manipulating the temperature we can grow certain strains and inhibit others. Other strains require specific nutrients to survive. As we saw in Chapter 6, wild-type cells, called *prototrophs*, can manufacture their own constituents from simple nutrients such as sugars, and an *auxotroph* is a mutant that can only grow when the medium is supplemented with another compound. Thus, *trp* mutants cannot make one of their amino acids, tryptophan, and can only grow if tryptophan is added to the growth medium. Other strains are resistant to antibiotics that kill wild-type bacteria; wild-type cells, str^s, are sensitive to streptomycin, and str^r mutants are resistant to it.

Bacteria can be handled easily with pipettes and sterile wire needles, so we can grow colonies on petri plates, where their characteristics can be determined. One of a geneticist's main tools is analyzing crosses between different strains, and this also can be done with bacteria, because even they have a kind of sex life. The discovery of sex and sexual interchange in bacteria is a fascinating story.

sex in *e. coli*

In 1946, Joshua Lederberg and Edward Tatum set out to find a way to use bacteria for genetic work. Only a few years earlier, Tatum had collaborated with George Beadle on the experiments with the mold *Neurospora* that produced the "one gene, one enzyme" principle, a fundamental insight of genetics. Lederberg and Tatum hoped to make greater progress by working with even simpler organisms. Luria and Delbrück had already shown that bacteria can mutate, just as other organisms do, so one could hope to find bacterial mutants to study. However, genetic experiments require crosses between different strains through some kind of sexual process, and bacteria were not known to have sex. Lederberg and Tatum reasoned that if bacteria did have sex, it would involve the fusion of different cells and this might provide an opportunity for recombination between two cells with different genotypes, just as in higher organisms. So if they could demonstrate recombination they would have indirectly shown that bacteria can engage in sex. They assumed that recombination in bacteria would be rare, so they designed an experiment to *select* any rare recombinants between two auxotrophic strains. One strain required threonine and leucine but could make its own methionine and thiamine – genotype *thr⁻ leu⁻ met⁺ thi⁺*; the other could make its own threonine and leucine but required methionine and thiamine – genotype *thr⁺ leu⁺ met⁻ thi⁻*. When these bacteria were mixed they produced a few *thr⁺ leu⁺ met⁺ thi⁺* prototrophs that could only have resulted from recombination.

Lederberg found that recombination occurs only between certain *fertile* strains, called *F⁺*, and other strains denoted *F⁻*, and only at quite a low frequency. Fertility in bacteria is like a disease, for when F⁻ cells are mixed with some F⁺ cells, they become converted to F⁺. This odd situation was clarified by the work of William Hayes in England, François Jacob and Elie Wollman in Paris, and L.L. Cavalli in Italy. First, they discovered that F⁺ cells contain a genetic factor, the *F factor* or *fertility factor*; F is a unique DNA molecule, as we shall see. F⁺ cells readily transfer copies of their F factor to other cells on contact, thus converting them to F⁺, and at a low frequency they also transfer some of their own genes to F⁻ cells, so recombination can occur. Hayes and Cavalli then discovered variant F⁺ strains called *Hfr*, for *high frequency of recombination*, which transfer their genes to F⁻ cells at quite a high rate.

The whole process of recombination became clear from crosses between a prototrophic Hfr strain that is sensitive to streptomycin and an F⁻ strain that is a multiple auxotroph resistant to the antibiotic, with a genotype such as *strʳ thr⁻ leu⁻ met⁻ lac⁻ gal⁻ thi⁻*; here each gene name signifies a metabolic deficiency such as the inability to synthesize threonine (*thr⁻*) or the inability to grow on lactose (*lac*). When the cells are mixed, the Hfr and F⁻ cells pair up and mate, a process called *conjugation*. After a short time, the cells are plated on media containing streptomycin, to kill the Hfrs, and one can test the F⁻ cells to see what wild-type genes they have acquired from the Hfr cells. In a classic experiment, Jacob and Wollman took samples of conjugating cells at various times, spun the cells in a blender to separate them, and plated them on various media to find out what genes had already been transferred from the Hfr. This experiment showed that each Hfr marker starts to appear at a specific time: gene *A* appears among recombinants only after four minutes, gene *B* only after seven minutes, *C* only after nine minutes, *D* only after fifteen minutes, and so on. Thus, it became clear that the Hfr transfers genes into the F⁻ cell in a *linear sequence*, the sequence in which they are linked in the bacterial chromosome.

Figure 10.1 explains these results. The F factor in an F^+ cell is a small circular DNA molecule. Contact with an F^- cell stimulates it to replicate itself and quickly transfer one copy into the F^- cell, converting it into an F^+. Occasionally during this process some F^+ genes get into the F^- cell, although it is still not clear just how this happens. However, in an Hfr cell the F factor has become physically *integrated* into the bacterial chromosome, which (as these experiments showed) is also circular, although much larger. Now upon contact with the F^-, the F factor again starts to transfer a copy

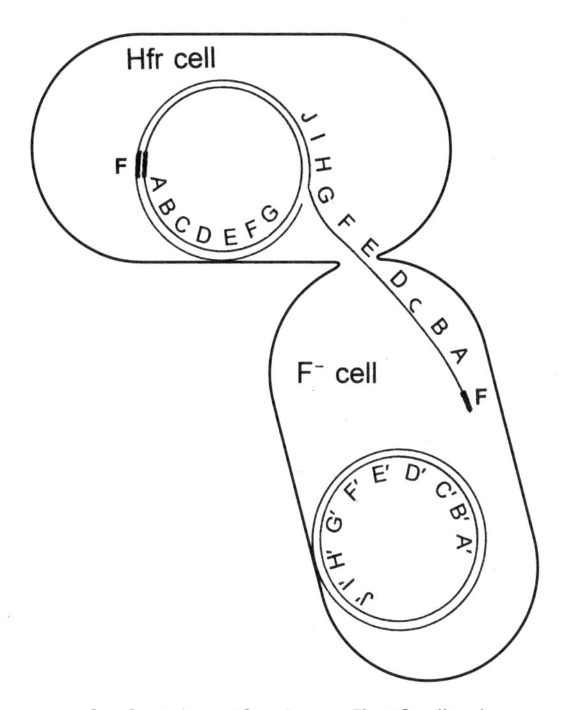

Fig. 10.1. *Transfer of DNA in an Hfr × F^- cross. The Hfr cell makes contact with the F^- cell, and a conjugation tube opens between them. The F factor, integrated into the Hfr chromosome, begins to replicate itself. But, since the F factor DNA is continuous with the rest of the chromosome, replication continues through the chromosomal DNA. So a copy of the Hfr chromosome is transferred into the F^- cell, with the genes entering in a linear sequence. Once the donor DNA is in the F^- cell, it can pair with the F^- chromosome, and recombination can integrate Hfr alleles into the F^- chromosome.*

of itself, but because the F factor and the bacterial chromosome are one piece, the copy contains bacterial genes, and these are transferred in their linkage sequence: *A*, *B*, *C*, and so on. Conjugation rarely lasts long enough (90–100 minutes) for replication to go all around the bacterial chromosome and pick up the tail end of the F factor, so the F^- cell is generally not converted to Hfr.

After conjugation, the newly transferred Hfr DNA recombines with the F^- chromosome, and various recombinants emerge. But instead of using frequency of recombination to create a general map of *E. coli*, with distances in map units, the units are minutes of transfer, and each marker is located by its time of entry. The F factor can integrate itself at many places and transfer in either direction from those points; each integration event creates a different Hfr strain with a unique origin, and these are used to map different parts of the genome. Refined genetic maps of *E. coli* (Figure 10.2) and of other bacteria have been created by combining mapping data from all these strains, along with another genetic method to be described later.

plasmids

The F factor is a prime example of a *plasmid*: a ring-shaped, self-replicating, extrachromosomal genetic element. Plasmids reside as passengers in cells and are replicated as the cellular DNA replicates. Being pieces of DNA, they can carry a number of genes, and they are very much like viruses; but plasmids do not form extracellular particles (virions) at any stage of their growth, as viruses do. Many other plasmids can transfer copies of themselves into other cells as F does, although some plasmids simply exist in their host cells and cannot spread autonomously from cell to cell. F is also a prime example of a subclass of plasmids called *episomes*, which can exist in their host cells as separate elements or can integrate themselves into their host's chromosomes.

resistance factors and antibiotic resistance

In 1955, a Japanese woman returning from Hong Kong came down with a form of dysentery, caused by bacteria of the genus *Shigella*.

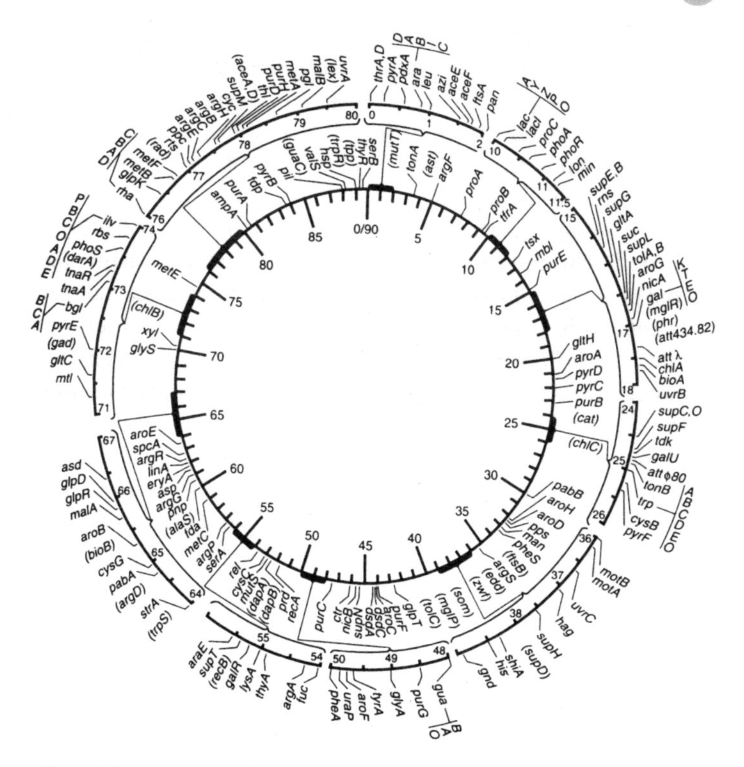

Fig. 10.2. *A map of the* E. coli *chromosome. The units are minutes of transfer in an Hfr × F⁻ mating. Selected portions of the map where many genes are known are expanded as arcs outside. The finer details of the map are determined by transduction, as explained below. This map was published several years ago, and the chromosome is now known in such detail that it would require a map covering several large pages.*

Shigella infections are usually controlled easily by antibiotics, but these bacteria were resistant to four different antibiotics: sulfanilamide, streptomycin, chloramphenicol, and tetracycline. This was very surprising, because antibiotic resistance is usually rare. If bacteria are plated on agar containing penicillin, perhaps one in 10^7 cells will be penicillin resistant. Similarly, one in 10^7 cells plated on streptomycin might survive. So if the bacteria were plated on a medium containing both antibiotics, we would expect only one in 10^{14} cells to be doubly resistant. Simultaneous resistance to four different drugs therefore seemed almost miraculous. Yet in the following years, dysentery in Japan became very intractable to drug treatment as cases of multiply resistant strains increased.

Tsumoto Watanabe found that multiple resistance is conferred by a plasmid called R (for resistance), which behaves much like F. Many types of R plasmids are now known that carry resistance to several drugs. R factors are remarkable, and dangerous, because they are infective: like F, they can transfer themselves from a resistant to a sensitive cell simply by physical contact, and they can often transfer into completely different species of bacteria. One study in Japan showed that the proportion of drug-resistant *Shigella* cells increased from 0.2 percent in 1953 to fifty-eight percent in 1965, mostly because of R factors. The worldwide situation has now become very serious, as Laurie Garrett documents in her book *The Coming Plague.*[1] Medical personnel increasingly encounter *Staphylococcus aureus* infections that are resistant to multiple antibiotics, including vancomycin and methicillin, which could formerly be used reliably to treat staph infections. Three species of bacteria that cause serious illnesses – *Enterococcus faecalis*, *Mycobacterium tuberculosis*, and *Pseudomonas aeroginosa* – are now becoming resistant to over a hundred antibiotics that have been used routinely against them. In general all kinds of infections that could be treated with antibiotics at one time are now becoming intractable.

The story of plasmids and antibiotic resistance is one of evolution (as well as human folly). Evolution depends upon natural selection, which means simply that in a given environment the most successful organisms are those whose genes make them best suited for life in that environment. By plating bacteria on agar containing streptomycin, for instance, we can select streptomycin-resistant mutants or cells carrying an R factor with a gene for resistance. Similarly, bacteria living in an environment laced with various antibiotics will be selected for resistance to them; R factors have become the most important agents of antibiotic resistance.

Plasmids have been found that confer resistance to a long list of antibiotics, including ampicillin, sulfonamides, chloramphenicol, tetracycline, kanamycin, neomycin, streptomycin, spectinomycin, and gentamicin, and to metals such as mercury, nickel, and cobalt. Some R factors can carry over ten different antibiotic-resistance genes at once. R-bearing bacteria have been reported in marine fish, and a comparison of bacteria in the River Stour in England sampled in 1970 and 1974 revealed that though the number of bacteria remained the same, antibiotic resistance had doubled. It is clear, then, that R factors are being strongly selected by widespread use of antibiotics. The rapid spread of R factors even across generic boundaries promises outbreaks of other diseases. Studies at the Birmingham Accident Hospital showed that drug-resistant *Pseudomonas* isolated from infected burn victims probably acquired R factors from the patients' own gut flora.

The greatest concern is that antibiotics are used extensively as growth promoters in feed for chickens, pigs, cattle, fish, and other livestock. Antibiotic production in the U.S. has grown from about two million pounds in 1954 to over fifty million pounds in 2002, and it is estimated that about half of this material is used on animals. Eliminating low-grade infections by such treatment helps animals gain weight. However, antibiotic treatment rapidly selects for drug resistance with consequent loss of growth-promoting effectiveness of the drugs. To keep up growth, antibiotic levels in feed have been doubled and tripled. As a result, the incidence of R-bearing cells in cattle and chickens has increased dramatically. Over seventy percent of the bacterial flora of most cattle intestines now carry R factors. In Great Britain, the potential hazard of R factors spreading from farm animals to bacteria in humans was documented in the Swann Report, which recommended a drastic cutback in the use of antibiotics as growth promoters. In 1998, the European Union banned the growth-promoting use of all antibiotics used to treat infections in humans, and responsible scientists and health advocates in the U.S. have petitioned the Food and Drug Administration to do the same. However, the economic returns of antibiotic sales to farmers are enormous – over 270 million dollars in the U.S. in 1983 – and in North America pharmaceutical companies have resisted the pressures to stop selling their products for feed to farm animals and have successfully fought legislation making such use illegal. Unfortunately, the U.S. government, even during quite liberal administrations, has sided with pharmaceutical and agricultural interests and has criticized the responsible actions of the European Union.

In Britain, E.S. Anderson studied R factors and has long been critical of the advertising policies of pharmaceutical companies. The promotional distortions are sometimes glaring, as in the Winthrop Laboratories' advertising for Negram:

> "Why give bacteria a 48-hour start?" said one advertisement, which carries a prominent statement that during the two days spent waiting for a sensitivity test on a bacterium isolated from a patient, one drug-resistant organism could produce 7.9231×10^{28} resistant bacteria. Any first class microbiology student knows that such limitless multiplication (which in this case would yield no less than 30,000 million tons of bacteria) does not happen in anything but highly contrived laboratory conditions, and certainly not in the body. Yet Winthrop chose to use this argument to recommend their drug to medical practitioners.[2]

Promotion of antibiotics in animal feed is often outrageous ("The best mouthful a pig ever had") and certainly irresponsible in promoting the liberal use of such strong selecting agents for R-bearing cells.

Although antibiotic manufacturers deny that their products cause any harm when used in animal feed, definitive demonstration of the hazards of using antibiotics in livestock came in 1984. In February 1983, Michael Osterholm of the Minnesota Health Department called the Centers for Disease Control (CDC) in Atlanta to report an unusual outbreak of gastrointestinal disease in the Minneapolis – St. Paul area, caused by *Salmonella newporti*. The bacteria turned out to be resistant to ampicillin, carbenicillin, and tetracycline, and all of them carried the same plasmid. Other infections by the same bacteria were reported in South Dakota by the state epidemiologist, Kenneth Senger. These bacteria were then traced to a feed lot that routinely added chlortetracycline to its animal feed; beef from the lot had been shipped to supermarkets where the Minnesota victims bought their meat. This episode contradicts the claims of the antibiotic and agriculture industries that their practices do not select for resistant bacteria and pose no threat to human health.

The hazards of R factors cannot be overstressed. They illustrate the folly of putting short-term benefits ahead of the possible long-term consequences. R-bearing strains of syphilis and gonorrhea now promise to add to the already heavy medical load of venereal disease.

Clearly, then, plasmids are powerful agents of genetic changes, changes that may be radical and unforeseeable in their pace and extent. As we shall see later, plasmids have also achieved a dual high profile of hope and fear in their use for genetic engineering.

lysogeny

Biologists experimenting with phages before the Second World War often claimed that some strains of bacteria harbor viruses, which appear occasionally and unpredictably in growing cultures. Scientists of the Delbrück school did not take these claims seriously and dismissed them as phage contamination owing to sloppy handling of bacteria. Then in 1950 André Lwoff and Antoinette Gutmann, in Paris, demonstrated that this phenomenon, called *lysogeny*, is real – that some bacteria do harbor an inactive form of a phage called a *prophage*. Lysogenic strains are immune to infection by the phages they carry. Occasionally the prophage in a lysogenic cell is spontaneously *induced*, so it emerges from its quiescent condition into an active, or *lytic*, state: its DNA replicates rapidly and the cell fills with phage particles and then lyses, as if it were infected by a phage such as T4. Lwoff later found that exposing lysogenic bacteria to ultraviolet light or various reagents could induce the lytic phase, so investigators can control phage growth.

Phages that lysogenize bacteria are called *temperate* in contrast to *virulent* phages such as T2 and T4 that can only grow lytically by multiplying and killing their host. When a temperate phage infects a sensitive (non-lysogenized) cell, it can either multiply lytically or it can become a prophage and convert the bacterium into a lysogenic cell. This cell can then grow into a clone of lysogenic cells, all carrying copies of the same prophage, because the prophage DNA is replicated as the bacterial chromosome replicates. Thus, temperate phage are very similar to plasmids. Like plasmids, they establish themselves as passengers in bacteria and are carried along as the cells reproduce.

In 1951, Esther Lederberg discovered that some strains of *E. coli* are lysogenic for a phage called lambda (λ), which has become one of the best understood and most useful viruses. Lambda provided the answer to the question of where the prophage is harbored within the cell. Mapping experiments showed that the lambda prophage is located at a specific site between the *gal* and *bio* genes (for metabolism of galactose and synthesis of the vitamin biotin). Allen Campbell then showed that the prophage is integrated at this site, just as the F factor is integrated in Hfr strains, so the prophage and the bacterial chromosome are a single, large DNA molecule. A newly infecting phage injects its DNA, which forms a circle. A specific attachment site of the lambda DNA has the same base sequence as a similar integration site on the *E. coli* chromosome between *gal* and *bio*:

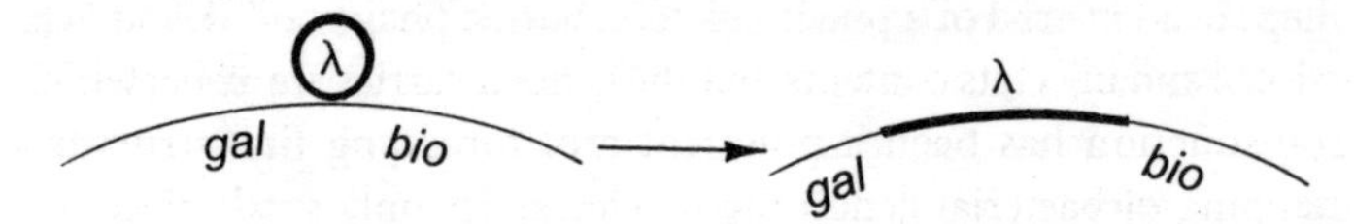

Through a crossover in this region, the lambda DNA becomes an integral part of the *E. coli* DNA. The integrated prophage has a control mechanism that keeps most prophage genes inactive, as well as the genes of any incoming viral genomes, so lysogenic cells are immune to infection by more phages of the same kind.

The prophages of other kinds of temperate phage, such as P1, do not become integrated into the host chromosome and simply remain as separate, plasmid-like elements. Some animal viruses can also exist in a kind of lysogenic state inside their host cells, a condition associated with some disease states. Many oncogenic viruses – those that induce tumors – also transform their host cells by integrating into the host's chromosomes.

gene transfer by virus

While trying to determine whether *Salmonella* could conjugate as *E. coli* does, Norton Zinder discovered that phages can transfer genes from one bacterial cell to another, a process called *transduction*. When a prophage is induced to go into a lytic cycle, the integrated prophage DNA comes out of the host chromosome by a reversal of the crossover that originally integrated the phage DNA into the chromosome. But sometimes this happens irregularly, so the phage DNA carries some bacterial genes on one side or the other of the integration site:

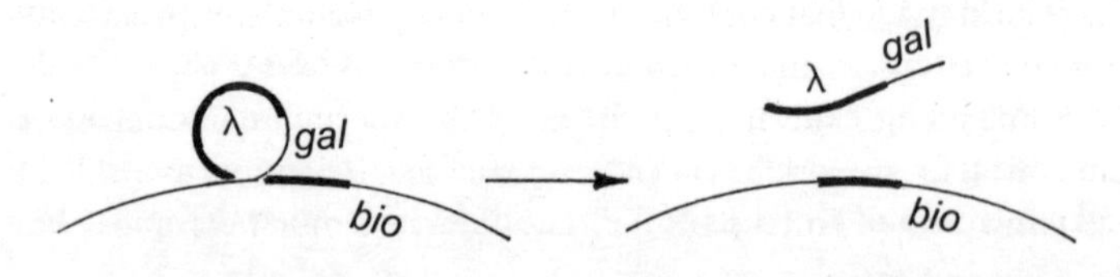

This piece of DNA can be incorporated in a phage particle, so phage lambda can carry the *gal* and *bio* genes into other cells. This is a *specialized* transduction, restricted to only a few genes. Phages such as P1 can perform a *generalized* transduction and carry *any* bacterial gene; the bacterial chromosome is chopped into smaller pieces during phage multiplication, and sometimes one of them is stuffed into a phage head instead of a phage genome. Such a phage can then attach to a cell and inject its contents, but the genes it carries are all bacterial. Transduction has been a powerful tool for doing fine-structure mapping of bacterial genes; the phage carries only small pieces of DNA, and an analysis of the way these pieces recombine with the new host chromosome can determine the sequence of bacterial markers.

transduction in humans

In 1955, Joshua Lederberg suggested that transducing viruses might be used to inject genes into human cells. At that time, the idea seemed like drastic speculation, yet now it is a real possibility, especially because of the recombinant-DNA methods described in Chapter 12.

Transduction could be used to (1) insert dominant "good" genes to correct a hereditary defect (phenotypically), (2) insert genes into crop plants to increase their utility, (3) alter bacteria to perform useful biological functions, and (4) induce genetic disease in a target organism. The principal question about transduction in mammals is whether animal viruses can pick up and insert genes into other cells. The answer is yes. For instance, when herpes simplex virus is used to infect a strain of mouse cells lacking a certain protein (thymidine kinase), 0.1 percent of the cells appear to have had the normal allele transduced. The normal allele could only have come from the viruses, and the transduced cell lines are quite stable and produce thymidine kinase for over eight months. In other model systems, adenoviruses and other animal viruses are being used to transfer genes, perhaps to cure genetic diseases. Transducing polyoma viruses have been made that contain mouse genes and can deliver the genes into nuclei of human embryo cells.

Some medical advances occur through experimental tests before an ethical framework has been developed for their application. That certainly was often true in the past, before so much attention was paid to ethical concerns, as when Banting and Best developed and tested crude preparations of insulin or when Christian Barnard electrified the world with an announced heart transplant. The same is true of tests of transduction in people. In 1958, Stanley Rogers was studying Shope papilloma virus, which induces skin lesions in rabbits. He found that infected cells had high levels of the enzyme arginase, which was postulated to be a product of a viral gene. Arginase degrades the amino acid arginine in blood serum. Rogers then found that a third of the lab technicians who had worked with Shope papilloma virus had serum arginine levels lower than ever reported in normal people. Evidently the virus had accidentally transduced the arginase gene into the technicians.

In 1970 Rogers learned of a family in Germany in which two sisters (later a third was born) suffered from hereditary hyperarginemia caused by a deficiency in arginase. In their blood and spinal fluid, the children had a very high level of arginine, which is associated with severe and incurable mental retardation, spastic paraplegia, and epileptic seizures. Rogers therefore injected Shope papilloma virus intravenously into the two girls (aged two and seven) and later into their five-month-old sister. The experiment failed to retard or correct the mental defects in the girls and was considered a failure. Nevertheless, this episode raises the question of how freely scientists should be experimenting with techniques that open far-reaching societal possibilities. Gene therapy was attempted in the face of considerable ignorance about the immediate consequences of injecting the virus, let alone any long-term effects. This is experimentation rather than gene therapy. Friedmann and Roblin[3] warn that such premature and unilateral experimentation might "serve as an impetus for other attempts in the near future" and therefore conclude by opposing any attempts in the foreseeable future until more information is obtained. Against this, Rogers argues,

> When one has a patient with a progressively deteriorating disease that is known not to respond to dietary or other known measures, and one has a possible means of stopping the progression of the disease with an agent that has been extensively investigated for 40 years, there appears to be little alternative other than to try it.[4]

However, although biologists seldom raise the issue, any technology for *correcting* human defects has the same potential for *inducing* defects in normal people.

In 1971, Carl Merrill and his associates reported the treatment of a human disease in cultured cells by transduction with phage lambda. They took skin cells of patients afflicted with galactosemia, an autosomal recessive condition characterized by inability to convert galactose into glucose (the sugar from which we derive energy). People suffering from this condition lack a function performed by one of the *gal* genes in *E. coli*. Upon infection of human galactosemic cells with lambda strains carrying bacterial gal^+ genes, Merrill and colleagues obtained cells with an even higher level of the required enzyme than normal cells have. These experiments are not critical proof that transduction took place, just as Rogers' experiments are not indubitable evidence. Nevertheless, taken at face value, Merrill's experiments do suggest that lambda can insert bacterial genes into human cells and that they can produce some tangible effect. As we shall see later, even more sophisticated techniques are now being used to transfer genes at will from one organism to another.

In summary, in this chapter we have seen how studying a seemingly esoteric system such as bacterial genetics can have great power in redirecting the thoughts and approaches of scientists. For example, bacterial plasmids and viruses appeared on the scientific horizon as agents that can act as gene *vectors*, which can carry extra pieces of DNA. Experiments with plasmids and viruses showed that they could replicate genes, shuttle them back and forth between different cells, and insert these extra pieces into the genomes of recipient cells. These are the key processes that have been adapted for DNA cloning and genetic engineering, as discussed in Chapter 12.

SELECTING MUTANTS

Selecting bacterial mutants, such as *ts* or auxotrophic strains, generally begins with growing some bacteria and exposing them to a *mutagen*, a substance or an agent (such as ultraviolet light) that produces mutations. If we wanted a *ts* mutant, we would plate the cells on agar medium and grow them at low temperature (28°C) until a number of colonies have developed. Then we

(continued overleaf)

(continued)

stretch a piece of sterile velvet over the end of a cylindrical block about the same diameter as the plate, and press it gently onto the culture plate. You can see why velvet picks up lint so quickly by looking at it under a microscope: velvet is like a jungle of needles. So when the velvet touches the plate, it picks up some cells from each colony. Pressing the velvet onto a fresh plate transfers some of these cells onto the new medium, thereby producing a *replica* plate that is an exact copy of the original. The replica is then grown at 42°C, and the plates are compared:

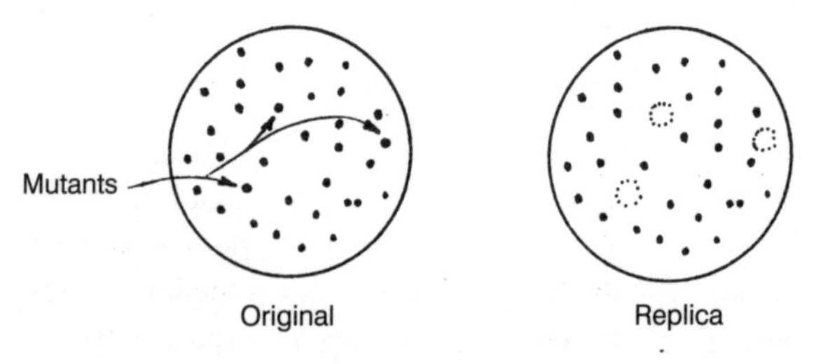

Any bacteria that do not grow on the replica apparently have a *ts* mutation, and we can grow them for further study. Using this *replica plating* technique, a master plate can be copied many times and the copies tested under different conditions. For instance, the original plate may contain all the amino acids the bacteria need, and a replica may be made onto a plate that lacks some amino acid, such as tryptophan. Then bacteria that do not grow on the replica plate are probably *trp* auxotrophs.

chapter eleven

gene regulation and development

Over the chapters of this book, just as during history, our conception of a gene has changed from a vague "factor" that is inherited in some regular way to a specific sequence of nucleotides in a DNA molecule that specifies the structure of a protein (or sometimes an RNA). Each gene is located somewhere in a chromosome, a part of the genome that governs the activities of an organism. But you may be inclined to complain, "Okay, I understand all this, and I see that a genome carries instructions for making a lot of enzymes and other proteins. But look at *me*! I'm more than a bag of enzymes, or even many bags of enzymes attached to some bones. If my genome really determines how I've operated all these years, it must have determined how I grew from a single-cell zygote into this big, complicated, highly organized structure made of many different kinds of cells. I want to know how my genes made me grow to be what I am!" So do we all. If you have thought anything like this, you have followed the thinking of many modern geneticists, who have devoted their professional lives to asking precisely these questions about the mechanisms of development.

The complaint above refers to "many different kinds of cells," and that phrase is worth looking at. Many books on biology and anatomy will show how many quite distinct types of cells compose the human body, or even the body of a much simpler animal (or plant, for that matter). Many of our cells are rather regular cubes or cylinders, which form large organs such as the liver, tubes such as the intestinal tract and blood vessels, or the lower layers of skin. Flattened cells, rather like paving blocks, fit together to make the linings of some tubes or the surface of the skin. The cells of our muscles are either very long cylinders or small spindles, which contain highly organized arrays of proteins that pull on one another to exert forces. Our nervous systems contain cells with long, thin processes – some of them meters long – that carry messages rapidly throughout the body. We are made of at least a hundred distinct types of cells, and the number grows as we learn more about their specializations.

So we can partially rephrase the question about how we develop into such complicated creatures: "How did all these cells become different from one another?" And the answer to that question is, in large part, "Each cell is made of distinctive proteins." When a cell makes a protein encoded by some gene, we say the gene is being *expressed*, so we can again rephrase our question: "How are genes regulated so they are expressed only at the right time and place?" That is what we shall explore in this chapter.

bacterial gene regulation

As before, we begin with simple systems in bacteria where the questions were first asked and answered. This work was carried out primarily in the 1950s and 1960s, much of it at the Pasteur Institute in Paris by François Jacob, Jacques Monod, and several American investigators who came to Paris to collaborate with them. The early work centered on an interesting system in *E. coli*, which lives in the mammalian gut. Mammals, especially when very young, have milk in their diet, and the principal sugar of milk is lactose. It therefore makes sense – since the bacteria have become well adapted to their environment through long evolution – for *E. coli* to have enzymes for metabolizing lactose; but, since the bacteria may encounter lactose occasionally but not all the time, it also makes sense for them to have a regulatory mechanism so they only make the necessary enzymes in the presence of lactose. Indeed, *E. coli* is a well-adapted organism, and the use of lactose is well regulated. Lactose is a double sugar (disaccharide) made of the simple sugars galactose and glucose bonded together; as a first step in its metabolism, the enzyme β-galactosidase splits the double sugar into its constituents, which the cells can easily metabolize further. If we grow *E. coli* in a medium containing no lactose, they make only a minute amount of β-galactosidase. However, if we add lactose to the culture, within three to five minutes the cells undergo a remarkable change: they start to make the enzyme about a thousand times faster than before, and this one enzyme can constitute a few percent of the cells' dry mass. Remove the lactose suddenly (by filtering or centrifugation) and the enzyme production falls to its initial level in minutes.

To investigate this system, Monod and his colleagues used the now classical method of mutational analysis. They looked for mutants that cannot metabolize lactose and found some that have a defective β-galactosidase, which they called *lacZ* mutants. They found other mutants that can make a perfectly good enzyme but still cannot grow on lactose; these mutants turned out to be defective in a protein called galactoside permease, a protein that transports lactose across the cell membrane into the cell. Mutants with a defective permease are called *lacY*. Mapping techniques showed that the *lacZ* and *lacY* genes, defined by these mutants, lie right next to each other.

The most interesting mutants, however, have a defect in their regulatory system, so they either cannot express the *lac* genes or else cannot stop expressing them. And indeed, mutants designated *lacI* make both β-galactosidase and the permease all the time, with no control. The *lacI* mutants, interestingly, map close to the *Z* and *Y* genes.

Experiments with these mutants revealed how the control system operates. First, it must be clear that expressing a gene fundamentally means transcribing it – synthesizing a messenger RNA from the gene. Remember that transcription is carried out by a large enzyme, the RNA polymerase, which only begins to transcribe at a specific sequence adjacent to the coding region, a sequence called a *promoter.* The promoter for the *lacZ* and *lacY* genes is a short region between *I* and *Z*. The direction of transcription – the direction in which the RNA polymerase moves to make an RNA transcript – is *downstream*; the opposite direction is *upstream*. So the promoter is just upstream of the *lacZ* gene. The *lacI* gene – now called a *regulator gene* – encodes a protein called the *lac repressor*. This is an allosteric protein (explained in the Box, pp. 184–86), which has two binding sites and can therefore bind to two different molecules. One site is specific for a short stretch of DNA, the *operator*, between the promoter and the *lacZ* gene; in the absence of lactose, the repressor binds to the operator, blocking transcription so the *Z* and *Y* genes are not expressed.

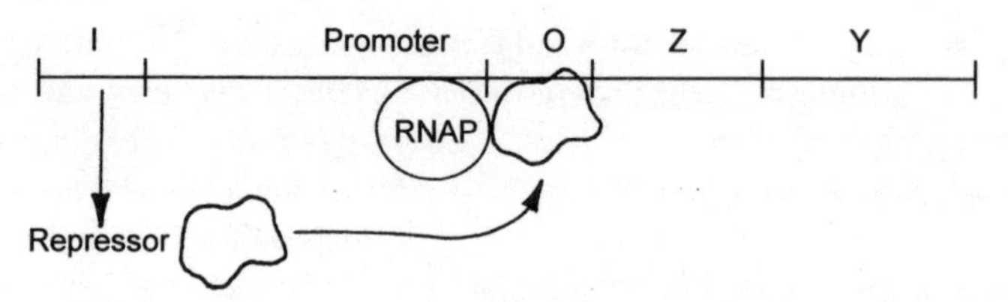

However, the repressor protein also has a binding site for lactose; when lactose is present, it binds to this site and changes the shape of the repressor subtly so it no longer has any affinity for the operator. The repressor thus comes off the operator, allowing transcription of the *Z* and *Y* genes to continue.
The *Z* and *Y* genes are thus expressed coordinately. These genes, being controlled by a single operator, constitute an *operon.*

The bacterial genome is highly organized into many kinds of operons, which are regulated in different ways. For instance, biosynthetic genes – genes encoding enzymes that synthesize cellular components such as amino acids – are regulated by a different logic, essentially the opposite of the *lac* repressor. Suppose a cell finds itself in a rich environment full of amino acids; if it is well regulated (and it is), it should stop wasting energy and materials needed to make its own. Genes for biosynthetic enzymes are organized into operons, regulated by other repressor proteins, but these proteins bind to their operators – and thus block transcription – only when there is an excess of the amino acid in question. For example, the enzymes needed to make histidine are encoded by a large block of genes, regulated by a single operator and repressor; this repressor binds to its operator, to shut off gene transcription, only when there is an excess of histidine in the cell. If the level of histidine falls, histidine molecules will come off the repressor molecules, the repressors will no longer bind to the operator, and the operon will again be transcribed.

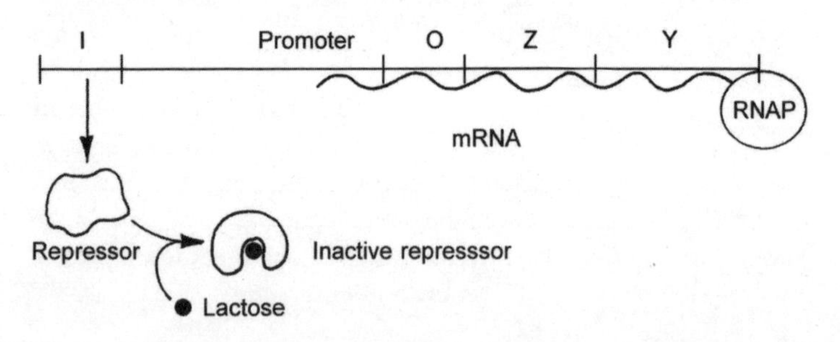

PROTEINS THAT BIND

We have talked a lot about proteins as enzymes, which catalyze chemical reactions, and we have mentioned proteins that perform other jobs. Understanding biology depends on the concept that the function of many proteins is to bind to other molecules called *ligands,* either permanently or temporarily. The ligand can be another kind of protein, and we see permanent binding in complexes of proteins that hold cells in certain shapes; our red blood cells, for instance, have the form of fat, concave disks, and they are held in that shape by about half a dozen proteins that form an elaborate network just under the cell surface. We see temporary binding in hemoglobin, the red protein that fills these cells. Hemoglobin carries oxygen from the lungs to all the other tissues of the body; it does so because it has a little site (next to the heme group, with its iron atom) that is very attractive to an oxygen molecule. The oxygen binds to this site as the hemoglobin passes through the lungs and then tends to leave this site where there is little oxygen, in the tissues.

Receptors are important proteins. Their function is to receive specific ligands and generally pass on the information that those ligands are present. A receptor protein has a little site that has just the right shape and chemical structure to attract some specific molecule. Receptors in the tongue and the nasal passages bind molecules from food or from the air, giving us the sensations of taste and smell. Once stimulated, the receptor cells send signals to the brain in the same way. The nervous system is made of many neurons, sometimes very elongated cells that connect to one another like wires in a telephone system. Each cell signals the next one in line by secreting signal molecules, which bind to receptors on the neighboring neuron and stimulate it to fire off a signal to the next neuron. Similarly, bacteria have receptors that recognize specific amino acids; when the receptors of a bacterial cell detect sugars or amino acids, they may signal the cell to swim toward the source.

These proteins all work because they are flexible. They change their shapes very subtly when bound to their ligands, and it is that slight change in shape that causes the protein to perform its next function – to stimulate the swimming apparatus of the bacterium or to cause a neuron to fire off an electrical signal. A hemoglobin molecule has slightly different shapes when it is bound to oxygen or not bound. This flexibility takes on special significance in *allosteric proteins*; these have two distinct binding sites that are specific for different ligands, and their shapes depend on which ligands are bound to them. A classical allosteric enzyme has an active site for its substrate and a second site for a regulatory ligand:

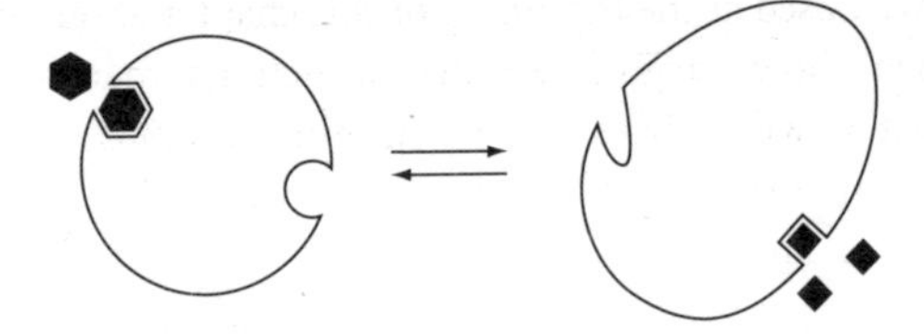

This kind of enzyme is often the first one in a specific biosynthetic pathway, and it can regulate the pathway through feedback inhibition. The regulatory ligand in this case is the end-product of the pathway, and if there is ever too much of this end-product the excess molecules will bind to the enzyme molecules, changing them slightly to inhibit their enzymatic activity. Then, when the level of the product falls, these molecules will fall off the enzymes and they will recover their activity. This regulatory mechanism ensures that the cell produces enough of each compound it needs but that it does not waste energy by producing a compound unnecessarily.

regulating eucaryotic genes

The whole question of gene regulation in eucaryotes contrasts markedly with that of procaryotes because typical eucaryotes live such different lives. Procaryotes are simple bacteria living in environments that may be constantly changing; the complex regulatory mechanisms described above have evolved because it is advantageous for bacteria to respond quickly, and genetically, to the arrival or disappearance of potential nutrients or cellular components. There are many eucaryotic microorganisms whose lives are similar, but most eucaryotes are plants or animals. Their cells live among other cells of the same organism, in environments that often change very little; we humans, for instance, have control systems (through our nervous and hormonal systems) to maintain the composition of our blood and other tissues as constant as possible, to stabilize our temperature, blood pressure, and other variables. Some cells, as in the liver, need to respond to changing situations, but most cells change very little. They have grown as complexes of specific proteins, which give them a distinctive form and function. Therefore, the most important questions about gene regulation in a plant or animal center on the question we raised at the beginning of this chapter about embryonic development: how a zygote becomes a functional mature organism. This is what we will address next, using animal examples.

Regulation of the bacterial type, with repressors and operators, is a good general model to keep in mind because it shows the general principle: genes are regulated by means of unique proteins that can bind to the DNA only at specific regulatory sites. However, the details in eucaryotic cells are quite different. Each gene is generally regulated by itself, not as part of a block of genes such as an operon. Each gene has its own promoter and is regulated by a complex of proteins that can bind to the promoter and to one another. This regulation can become unimaginatively complicated. Eucaryotic cells contain general proteins that can bind to all promoters and initiate transcription; then they have a variety of other proteins that are more or less specific for different genes or classes of genes. All these proteins pile up on one another at a promoter site, and it is only when they are all in place that RNA polymerase molecules can bind to that promoter and begin to transcribe the gene. There is little point in describing any one gene and the proteins that bind to its promoter, because it would only be listing a series of otherwise meaningless names. The take-home lesson is just that the decision to express a particular gene, during embryonic development or later, is made through interactions of several regulatory proteins.

embryonic development in general

An embryo develops from a single cell – a zygote – into a complex of many specialized cells. A zygote is *totipotent*; that is, through successive division it has the potential to become any kind of specialized cell such as a skin or brain cell. During development, most cells that specialize lose their totipotency, and no amount of experimental tweaking can cause them to back up and regain it. However, *some* cells can be tweaked in this way. For example, the developed cell of a mammary gland provided the totipotent nucleus for the zygote that became the cloned sheep Dolly. The body does contain generally totipotent cells called *stem cells*, which are used to replace mature cells as they die and are lost. Stem cells are useful for various kinds of therapeutic treatments in humans. Obtaining stem cells is difficult, however. A source of easily obtained stem cells is early human embryos; such embryos can be disrupted, and every cell becomes a totipotent embryonic stem cell. The ethics of using embryos for this purpose has been the center of a stormy debate in recent years.

In discussing embryonic development, we must distinguish two events. First, sometime during development, each cell is *determined* to become a particular type; this is the process in which a molecular decision is made and a cell's fate is at least partially set. However, cells in the embryo that are determined to eventually become quite different may still look essentially alike; it is only later that a cell *differentiates* and actually acquires its distinctive form and its complex of distinctive proteins.

We cannot yet describe the whole sequence of events that turns an animal zygote into a multicellular animal such as a human. Classical embryology provided detailed descriptions of the process, with little explanation of how it happens. It is important to see that a zygote begins to divide into cells: into two, then four, then eight, and eventually into a ball of many cells. This becomes a hollow ball, a *blastula*, with all the cells on the outside, and then a massive cellular movement occurs in which many cells migrate into the inside of the ball to form a *gastrula*. After some additional movements and reorganization, the embryo is divisible into three broad layers: an outer *ectoderm*, which will form skin and the nervous system; an inner *endoderm*, which will form mostly the lining of the intestine and organs attached to it; and a *mesoderm* between these layers, which will form the bulk of the internal organs. Although some genes responsible for this degree of differentiation have been identified, on the whole we cannot give useful examples of gene action during these early events. However, investigators have learned a lot about some limited developmental sequences, even to the point of identifying specific genes that produce them. Examining a few of these cases will provide a feeling for the ways genes can interact to produce structures.

A cell may be directed to differentiate into a particular type by either internal or external instructions. External instruction, as demonstrated by classical embryologists, has been called *induction*. As an embryo develops, cells engage in massive movements from place to place, bringing cells of different types into contact with one another. When this happens, cells of one type may send instructions to cells of the other type. One instance of induction happens in eye formation in the vertebrate head, as seen in experiments with chick embryos (Figure 11.1). The core of the central nervous system, which forms quite early, is a tube lying just under the back of the embryo, enlarged in the head into what will become the brain. The eyes begin to form as cuplike outgrowths from the brain; these cups have no lenses, and they instruct some of the outer ectoderm cells lying over them to become thicker and form lens tissue. If a cup is removed by microsurgery, a lens does not form where it normally would; if the cup is moved to another location, it induces formation of a lens there.

Induction clearly depends on some cells – the inducing tissue – already being determined and partly differentiated, so we cannot

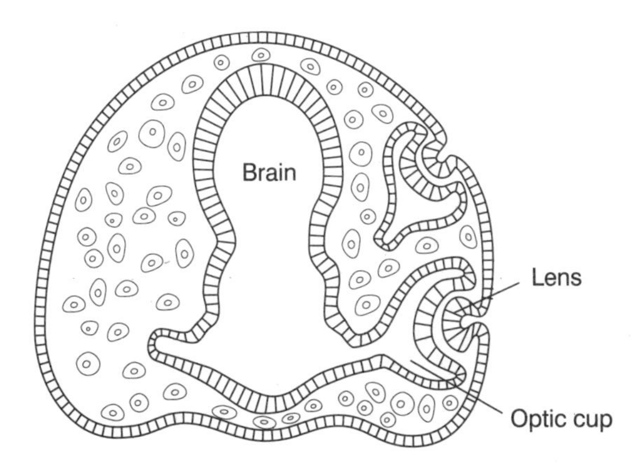

Fig. 11.1. *A chick embryo's eye begins to form as a cuplike outgrowth from the brain. This cup induces the tissue over it to form a lens. If a cup is removed, no lens forms. If the cup is transplanted to another location, it induces a lens there.*

explain all differentiation by inductive interactions. Now we must back up and ask how cells may be determined through internal mechanisms. We know examples of two general mechanisms, where differentiation is governed by *time* or by *position.*

regulation by time in a chick's wing

A clear example of a clock mechanism has been demonstrated in the development of the chick wing (Figure 11.2). The wing grows from an embryonic limb bud, made of a core of mesoderm covered by a layer of ectoderm, including an *apical ridge* at the growing tip of the bud. The cells of this ridge instruct the underlying mesoderm, and they appear to have an internal clock that determines what instructions they will give. J.H. Lewis, D. Summerbell, and Lewis Wolpert removed ridges of various ages and grafted them onto buds

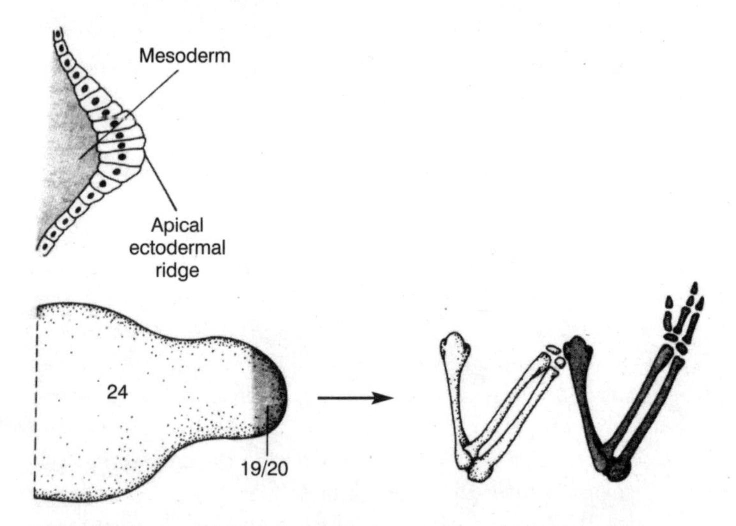

Fig. 11.2. *A normal chick wing contains a typical set of vertebrate limb bones: a humerus (upper arm), radius and ulna (lower arm), and a set of wrist bones and phalanges (fingers). The mesoderm of the limb bud is instructed by the apical ridge to lay down each of these bones in turn. If a young ridge is grafted to a limb that has already started to form the more basal bones, the ridge instructs the bud to repeat these structures because its internal clock is still at an earlier stage.*

of other ages. If they grafted a young ridge onto a bud that had already laid down the humerus and radius–ulna, the ridge instructed the tissue to repeat these bones. However, if they grafted an older ridge onto a young bud, it only instructed the bud to lay down the terminal phalanges.

This experiment indicates that a clock in the apical ridge cells first turns on genes that specify the humerus; then the clock ticks, those genes are turned off, and other genes are expressed that specify the radius and ulna. The clock ticks again, and another set of genes is expressed that directs the formation of the phalanges. We do not know the mechanism of such a clock, but it takes time to perform each step in a process in which regulatory proteins form and then bind to particular genes and induce the formation of other proteins. It is not hard to draw theoretical circuits that create a clocklike mechanism.

determination by position in a fly's body

Fruit flies have been as valuable for studying the genetics of development as they were for classical studies of inheritance. The most obvious way to determine that certain cells should differentiate in a particular way is by their location. Eyes and a mouth should develop on the head, legs on the bottom (ventral) side of the body in the middle, and wings on the top (dorsal) side in the middle. But how can a place be specified genetically? If an egg cell were perfectly uniform and symmetrical, it would probably be impossible to give it positional information and make it develop properly. However, an egg is apparently given initial positional information by the cells around it as it develops. A *Drosophila* egg develops in a fly's ovary surrounded by fifteen other cells, called nurse cells, which make the ends of the egg different. They are connected to the developing egg by small channels, through which they can inject materials. Among these materials are some specific mRNAs, which start the differentiation of the egg (after it is fertilized and becomes a zygote). The logic of initial determination is shown in Figure 11.3. The genes involved are named, as usual, by mutants for those genes, which sometimes have funny names. Gene names are italicized; the names of the proteins they encode are capitalized and printed in roman type (in the caption, italic and roman are reversed).

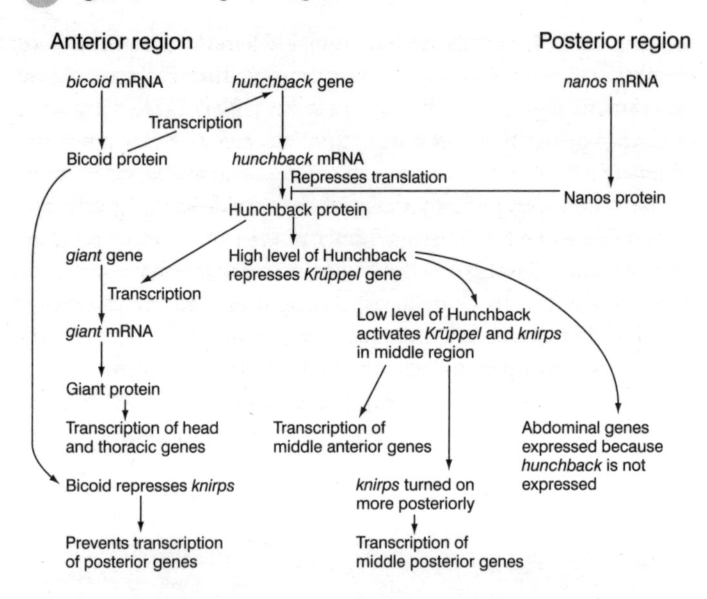

Fig. 11.3. *Initial interactions among genes determine the anterior–posterior axis of a fruit fly embryo. In the anterior end, the* bicoid *mRNA has been injected, and the Bicoid protein causes transcription of the* hunchback *gene. This leads eventually to the transcription of head and thorax genes, and repression of abdominal genes. In the posterior region, the* nanos *mRNA has been injected; the Nanos protein represses formation of the Hunchback protein. Therefore, abdominal genes are expressed because Hunchback is not present.*

The critical initial events are injection of *bicoid* mRNA into the end that will become the anterior (head) end and of *nanos* mRNA into the future posterior (tail) end. The protein translated from each of these messengers then either induces or represses formation of at least one other protein. Each sequence of events then becomes amplified, step by step, until it leads to the formation of distinctive structures from distinctive proteins.

Once the initial anterior–posterior axis has been established, a series of other genes are turned on that eventually divide the growing embryo into segments, for the fly's body is organized in this way: into five head segments, three in the thorax, and eleven in the abdomen. The proteins involved in this process can be identified and located by making antibodies against them and attaching dyes to these antibodies; when a developing embryo is immersed in a solution of these antibodies, they attach wherever the proteins are located, creating bands of dyes across the embryo. These experiments have shown that first a series of *gap* genes are turned on; the proteins they encode turn on a series of *pair-rule* genes, which divide the embryo into fourteen segments. Next, some *segment-polarity* genes are turned on; they divide each segment into an anterior and a posterior half. Then a series of *homeotic genes* are induced; these fascinating genes begin to cause each segment to differentiate into its characteristic structures, and they were first identified in mutants with bizarre forms. For instance, *Antennapedia* mutants sport a pair of legs on their heads in the places where antennae ought to grow. By combining certain homeotic mutants, Ed Lewis was able to grow four-winged flies instead of the normal two-winged ones.

forming a fly's eye

A remarkable series of events, involving several genes, has been worked out in the formation of the compound eye of *Drosophila*. The insect's compound eye contains about eight hundred units called ommatidia. Each ommatidium contains a precise pattern of about twenty cells. These cells form a flat layer where they interact only with their neighbors; there are no cells above or below. The default setting of these cells is to become lens cells; but if they engage in a series of genetically programmed interactions with one another, they begin to differentiate into photoreceptors, the cells that receive light and send signals to the brain. Differentiation in each ommatidium happens in a defined sequence (Figure 11.4). First the central photoreceptor, R8, differentiates. R8 appears to induce its neighbors, R2 and R5. These, in turn, induce R3 and R4 on one side, R1 and R6 on the other. Finally, R7 is induced. The remaining cells in a group of twenty become lens and other structures.

Differentiation begins when a wave, seen as a visible groove in the cell layer, passes across the eye from posterior to anterior. As the wave passes, cells begin to divide and then stop early in the cell cycle. The orientation of the wave – and thus of the photoreceptor clusters – depends on expression of the *wingless* (*wg*) gene, whose product is a protein of a general signalling pathway, a series of proteins that carry

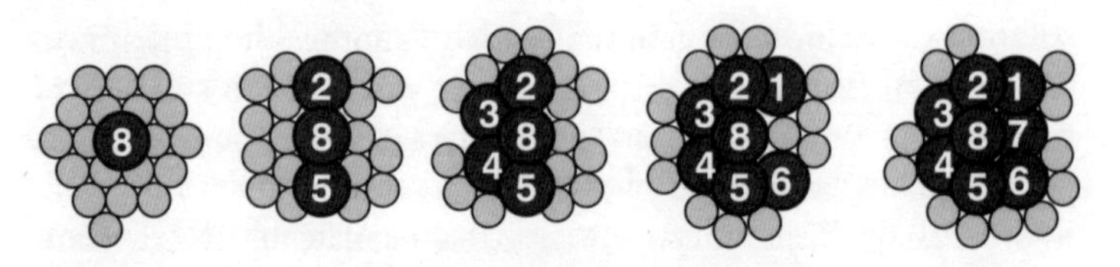

Fig. 11.4. *The eight cells that become photoreceptors in each ommatidium of the fruit fly eye differentiate in a specific pattern because of the interactions among several genes and their protein products.*

specific instructions for differentiation. The cells just behind the wave then start to differentiate and to secrete the Hedgehog protein (Hh). Hh diffuses to undifferentiated cells and stimulates them to secrete Decapentaplegic protein (Dpp). Dpp diffuses to more anterior cells and induces them to start differentiating and to release Hh, which stimulates another round of Dpp production. This cycling of Hh and Dpp proteins causes the wave to continue propagating.

To understand how the next cells are determined, we begin by looking at cell R7, even though R7 is the last cell to differentiate. A mutant, *sevenless* (*sev*), defines a gene that is required for the differentiation of the R7 cell. R7 is distinctive because it is a receptor for ultraviolet light, and *sevenless* mutants can be detected and selected by means of this deficiency. The Sevenless protein is a receptor on the surface of the cell that will become R7, and the ligand it binds to is another protein, called Boss, encoded by the *bride-of-sevenless* (*boss*) gene and expressed on the surface of R8. So the Boss protein on R8 instructs its neighboring cell to develop into R7.

However, the Sevenless protein is also expressed on the surfaces of the R3, R4, R1, and R6 cells. So why don't they become R7 cells instead? It is because these four cells express the *seven-up* (*sup*) gene, and the Seven-up protein restricts them to the fate of becoming just what they are. An additional mutant, *rough*, reveals more about the interactions among these cells. The *rough* gene is normally turned on in R2 and R5, and it allows R2 and R5 to induce the formation of R3 and R4, on one side, and R1 and R6, on the other side.

We don't yet know just how all these interactions are programmed, but analysis of these few mutants has revealed a pattern: one cell makes certain proteins, which then interact with proteins in adjacent cells to turn on one set of genes and turn off another set. Returning to the question that started this inquiry, we can point to a variety of mechanisms through which genes interact. Some genes encode proteins that directly bind to other genes, turning them on or off. Another class of genes encode proteins that regulate the translation of specific mRNAs into their proteins. Still other genes encode proteins that diffuse from one cell to another, inducing them to change or inhibiting their changes, and the products of some genes are proteins that interact with one another on the surfaces of adjacent cells. The logic of gene expression during a complex process such as embryonic development resides in networks of interactions that may entail all of these possibilities and others that we have not mentioned.

chapter twelve

dna manipulation: the return of epimetheus?

In Greek mythology, the Titans were a race of giants who stood between the gods and humans. The Titan Epimetheus was commissioned by the gods to create the plants and animals and distribute various attributes and characteristics among them. But Epimetheus – whose name means "afterthought" – was given to acting rashly, and he ran out of attributes before he got to humankind. Therefore he had to ask his brother Prometheus to steal the gift of fire from the gods to give it to humans. This incurred the wrath of Zeus, who sent Epimetheus a beautiful wife, Pandora, whom Zeus knew would cause him grief. Pandora released the contents of a forbidden box (also sent by Zeus) that contained all the ills of the world. Today, some people might see the geneticist as a modern Epimetheus, whose thoughtless meddling with nature may unleash new horrors from some scientific Pandora's box. These fears of a modern Epimetheus have become particularly acute since geneticists have learned how to manipulate DNA directly. Now, indeed, the geneticist seems godlike, with an ability to create organisms that could never have existed without this new technology.

We have shown how, from the earliest days of farming, humans applied their crude knowledge of reproduction to breed plants or animals in order to combine desirable properties. In the twentieth century, applications of Mendel's laws of heredity led to great improvements in the efficiency of this traditional type of breeding. But this type of improvement could be accomplished only by combining desirable features *within* a species. With the new DNA technology, features of *any* organisms could be combined in one, to make creatures reminiscent of the Chimera, a fire-breathing monster of Greek mythology that had a lion's body, a goat's head, and a serpent's tail, and which could be created only by the gods. It is this type of DNA manipulation that we review in this chapter. Some of the social and ethical implications of this manipulation are covered in the next chapter.

recombinant dna and restriction enzymes

Around 1972, Annie Chang, Paul Berg, and Seymour Cohen realized that restriction enzymes could be used for recombining virtually any two pieces of DNA to make what is termed *recombinant DNA*. This is the basis of the new technology that has revolutionized research in genetics and molecular biology. The method depends on the fact that complementary single-stranded nucleic acid molecules will tend to bind to each other with great precision. Hence, in a solution, a molecule with a sequence such as GCTAT will find and bind to a molecule with the sequence CGATA. DNA cut by restriction enzymes is particularly important. Remember that most of these enzymes cut palindromic sequences symmetrically, generally leaving short, complementary, single-stranded ends of two to four bases. For instance, *Sal*I cuts like this:

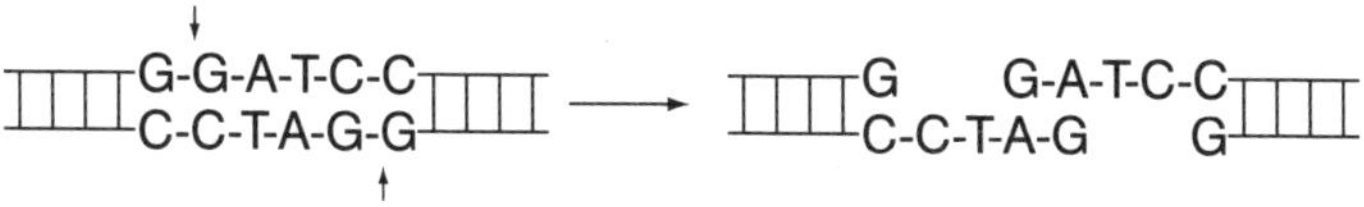

```
    ↓
G-G-A-T-C-C              G      G-A-T-C-C
C-C-T-A-G-G     →        C-C-T-A-G      G
          ↑
```

All DNA molecules that have been cut with the same restriction endonuclease have identical single-stranded sequences, which can hydrogen-bond to each other, even though they are very short. Hence, if we cut DNAs from two different organisms with the same restriction enzyme and then mix them together, we should obtain many recombinant molecules made of segments of DNA from both species.

In practice, research is always focused on the DNA from a particular *donor* organism, perhaps a mouse or even a human. This is the DNA we would like to isolate and manipulate experimentally. It is easy to extract the total genomic DNA from a donor organism, simply by grinding up cells and precipitating the DNA with ethanol. The donor DNA is then treated with a restriction enzyme, to make small, easily manipulated fragments, and these fragments are inserted into small replicating DNA molecules that act as *vectors*. In Greek, *vector* means "carrier," and a vector molecule carries the DNA of interest so an experimenter can manipulate it. Vectors can replicate, so they can amplify an inserted donor fragment, providing many copies for analysis. The most commonly used vectors are *plasmids*, which we introduced in Chapter 10: small, circular DNA molecules found in bacterial cells, which often carry genes for bacterial resistance to antibiotics. Plasmids are not part of the main bacterial genome, so both plasmid-bearing and plasmid-free strains are available. Because of its small size and circularity, plasmid DNA is quite easy to separate from bacterial genomic DNA. Certain small viral DNAs are also commonly used as DNA vectors.

After cutting the donor DNA in fragments with a restriction enzyme, we use the same enzyme to cut a vector molecule that has one target site for that restriction enzyme. The two DNAs are mixed and some recombinant DNA molecules form; these are vector molecules with a donor fragment inserted:

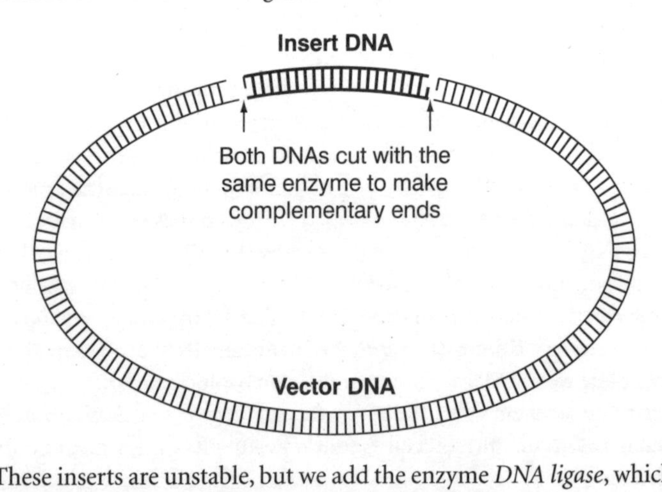

These inserts are unstable, but we add the enzyme *DNA ligase*, which joins the DNA backbones, completing and stabilizing the insertion.

The reaction mixture contains many vector molecules enclosing many different donor inserts. The next step is to introduce these recombinant molecules into living cells where they will be able to replicate. If plasmids are used as vectors, the recombinant-DNA mixture is simply added to plasmid-free bacterial cells, generally those of *E. coli*. When treated appropriately, the cells can pick up recombinant vectors through their cell walls and membranes. Generally only one molecule enters per cell. When the cells are plated, they will grow and divide, amplifying the plasmids and their inserts. Each of the resulting bacterial colonies is then a *DNA clone*, a population of identical bacterial cells amplifying a certain donor DNA fragment.

These colonies are collected and preserved. Together, they comprise a *genomic library*. If enough colonies are collected, there is a good chance that every part of the donor genome is represented at least once in the library. So we might end up with a library of mouse genomic DNA carried on plasmid vectors inside bacterial cells.

The next step depends very much on the goal of the experimental program. Some options are described in the next sections.

studies of individual cloned fragments

Our research often focuses on a specific donor gene that we want to study or use. With luck, the genomic library will contain a clone carrying this gene, but how can we find it? Many ingenious ways have been developed to find a particular gene needle in the genomic haystack. If the library has been made from wild-type genomic DNA, we can add the DNA from each clone to some organisms that are mutant for the gene of interest. Then we screen these cells to look for any that have been converted to the wild-type phenotype. This method, called *functional complementation*, works because cells of most organisms will take up DNA. The DNA may be taken up passively, or its entry is encouraged with an electric current. If the recipient cells are large enough, DNA can be injected. There are even "gene guns" that shoot DNA into a cell aboard metal particles. In each case, the introduced DNA may insert somewhere in the recipient cell's genome, taking up residence as a normal gene. If the cloned DNA we have introduced carries the wild-type allele of the mutant gene in the recipients, they may be transformed to wild-type, thus identifying the clone that carries the gene of interest.

Once a gene has been identified as the one being sought, many possibilities open up. One obvious step is to sequence the DNA of the insert (review the Box, pp. 140–42) and deduce the amino acid sequence of its protein product. Vast numbers of protein sequences are available in databases now, and from these it is often possible to tell what kind of protein a gene encodes.

Once a gene of known function is available, it can be used to detect similar genes in related organisms, because there is considerable evolutionary conservation of sequence among different species. One way to find related genes is to use a powerful technique called the *polymerase chain reaction (PCR)*, invented by Kary Mullis in 1984. It depends on the fact that DNA replication must begin with a primer, a short piece of DNA (or RNA) that the DNA polymerase then extends. The sequence of the cloned gene is used to design short primers at each end of the segment of interest. These primers point at each other on opposite DNA strands, and the strands act as templates for DNA polymerization enzymes, which run rapidly back and forth between the primers, exponentially amplifying the segment between them (Figure 12.1). If an amplified fragment of suitable size is found, it can be sequenced and compared with the sequence in the original cloned DNA.

PCR is also useful in diagnosing human disease alleles. Most recessive alleles in a population are harbored unexpressed in heterozygotes, and the disease appears only in the homozygous offspring of heterozygous parents. So it is important for medical geneticists to be able to detect heterozygotes. If a specific disease allele has been characterized by cloning and sequencing, then the PCR primers can be designed to detect that allele even though it is accompanied by a wild-type allele in a heterozygote. Dominant disease alleles, such as Huntington's disease, sometimes do not show up until late in life. PCR can be used to detect the allele before it is expressed. Thus, cloning, sequencing, and associated techniques such as PCR have come to play a standard role in medical diagnostics.

Another important use of a cloned gene is as a *probe*, to detect and locate homologous segments of other nucleic acids by complementary base pairing. The DNA of the cloned gene is tagged with either radioactive atoms such as ^{32}P or a fluorescent chemical. This DNA is separated into two single strands, which can be used as

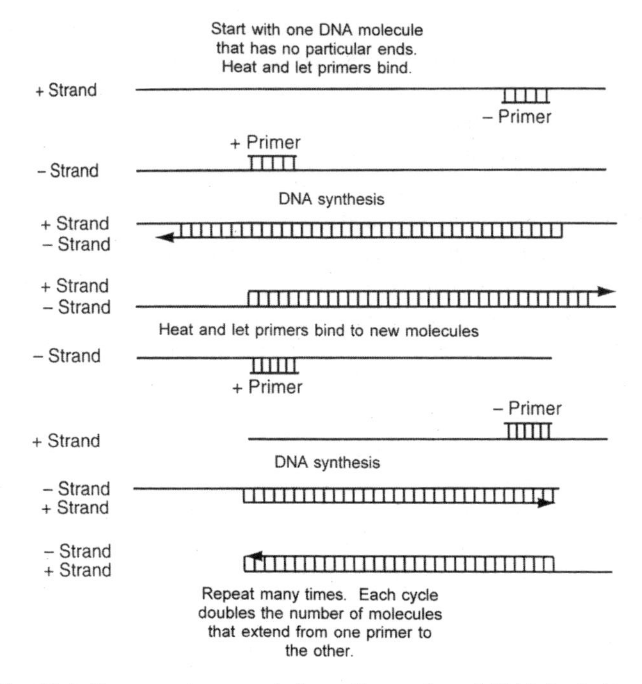

Fig. 12.1. *To run a polymerase chain reaction, a piece of DNA is heated to separate into single strands. Primers are added that bind to sites on the two strands, and the strands are replicated with an enzyme that is stable at high temperatures. The resulting molecules are again separated with heat, and the cycle is repeated, each time producing twice as many molecules as there were before.*

probes by adding them to various assemblages of other single-stranded nucleic acids (DNA or RNA). Because single-stranded nucleic acids with complementary sequences will bind to one another quite specifically, we can locate the tagged probes and thus locate a homologous segment. Probes are very versatile tools in genetics. They can be used to find the position of a gene on a chromosome or to determine the presence and the quantity of mRNAs transcribed by a specific gene under a variety of circumstances.

Both PCR and probing are used in forensics to detect individualistic DNA patterns that might help identify a criminal.

Bodily fluids such as blood and semen are convenient sources for DNA to be tested. Because of the sensitivity of PCR, even minute samples, such as the few cells adhering to a human hair, can give reliable results. The DNA is cut with a restriction enzyme, run on a gel, and treated with the probe. The outcome of such tests is a set of labelled DNA fragments in a graphic display called a *DNA fingerprint* (Figure 12.2). The tests are based on the discovery that although all humans have the same genes in the same positions, there is quite a lot of variation in the DNA that lies between genes, making for variations in the size of fragments cut with a restriction enzyme. Probes and PCR primers are available for particular variable regions, and they show that each person has a distinctive pattern of fragments. Then DNA from a sample obtained at a crime scene is run along with the DNA of suspects (and victims), making a set of parallel fingerprints; this test can at least show that a suspect is innocent and serves as strong evidence that one person is guilty.

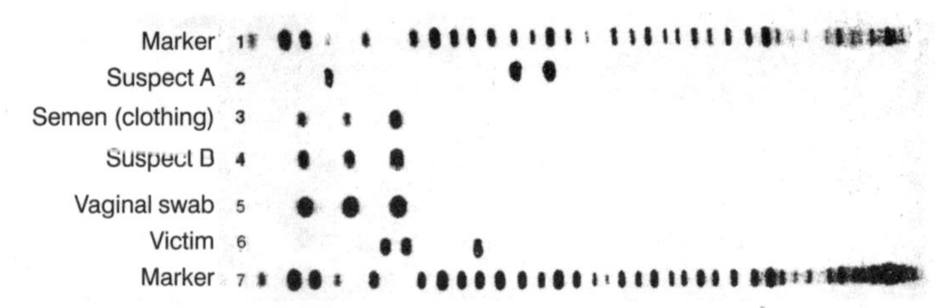

Fig. 12.2. *DNA fingerprinting is used to distinguish between two suspects who have been arrested in a case of rape. DNA samples from suspects A and B are run alongside samples from the victim (lane 6), a semen sample from her clothing (lane 3), and a vaginal swab (lane 5). The evidence clearly matches the sample from suspect B and excludes suspect A.*

transgenic organisms

One of the greatest applications of cloned genes is in making organisms that have been modified with foreign genes, especially for commerce. The genome of a plant or animal can be modified by adding some specific cloned DNA that is thought to improve the properties of that recipient organism. The cloned gene added is called a *transgene*, and the plant or animal modified by its addition is a *transgenic organism.* In the popular media this process has become known as "*genetic modification,*" a confusing phrase because traditional breeding practices dating back to the beginning of agriculture have also caused genetic modification. The more proper terms *genetically modified organisms* (*GMOs*) and *genetically modified food* (*GM food*) refer exclusively to transgenesis.

The significant difference between transgenesis and conventional breeding is that the transgenic DNA can come from *any* other organism, vastly increasing the range of options for modification. In conventional breeding techniques, a new allele must be introduced by breeding with organisms of the same species or very closely related species. If it were useful, transgenic methods could be used to transfer fish genes into plants or bacterial genes into mammals. So the possibilities for improving foods are limited only by the imagination – a fact that gives pause to many people who have considered some of the detrimental historical effects of the human imagination.

Transgenic modification is also appealing because of its speed. The new DNA is added in a matter of hours or days. The modified organism generally matures in weeks or months and then is ready for testing. Introducing a new gene by traditional breeding takes several generations, often up to a decade in all.

Transgenesis is a very versatile tool. It has the potential of making farming easier or more efficient and improving the quality of food. For example, pork or beef can be made leaner with less fat. The quantity of food produced per animal can be improved: each cow can produce more milk, each hen more eggs, or each wheat plant more grain. Some investigators are working on providing crops with their own genes for nitrogen fixation. Nitrogen, which makes up seventy-nine percent of the air, is a critical constituent of proteins and nucleic acids. However, plants can only use nitrogen that has been *fixed* – that is, reduced to molecules such as ammonia and nitrate. Ammonia is produced in the chemical reaction $3H_2 + N_2 \rightarrow 2NH_3$, primarily by nitrogen-fixing bacteria; these bacteria often grow in nodules on the roots of legumes such beans and peas or trees such as alder. Industrial processes can also combine nitrogen and hydrogen to get ammonia, but nitrogen fixation performed this way requires a phenomenal amount of energy. Furthermore, agriculture that depends on fertilizing plants with ammonia and other nitrogen compounds is far inferior to organic methods that nurture a natural soil ecosystem. Farmers have traditionally added nitrogen to their soil by growing a crop of nitrogen-fixing plants and then plowing them under to nurture another crop. But it might be desirable to give crops their own nitrogen-fixation (*nif*) genes. In the bacterium *Klebsiella pneumoniae*, the *nif* genes are clustered together, and these genes could be moved into cells of desirable food plants and expressed functionally.

Plant crops can also be made resistant to destructive pests such as fungi or insects. Insect damage causes large crop losses to farmers every year. Several corporations have experimented with adding genes for insect resistance to crops such as corn. They have commonly used Bt toxin, produced by various genes from the bacterium *Bacillus thuringensis*. Plants transgenically modified with these genes have their own insect defense system. Another approach is to modify plants with genes derived from certain bacteria to make them resistant to specific herbicides, notably the herbicide glyphosate, sold under the name Roundup. Farmers can then spray an entire field to kill the weeds but leave the crop plants unharmed. This makes farming much more efficient. But all these procedures are very controversial, as we explore in Chapter 13. They have important implications for human health, for the health of ecosystems, and for the business of agriculture.

Transgenic methods can improve the nutritional value of food plants; "golden rice," for instance, is a transgenic rice with genes that produce high levels of vitamin A. Genes for salt tolerance give a recipient plant the ability to grow in otherwise non-arable lands with high salt concentrations – for instance, tomato plants suitable for farming in certain regions in Israel. Some areas, such as those near the sea, are naturally salty. But soil salinity is a growing problem worldwide, because salt has accumulated in the soil of many areas as a result of excessive irrigation during farming. Producing salt-resistant crops typifies the efforts being made to increase the range of farming by increasing plant tolerance to heat, cold, and minerals.

Some investigators are trying to develop plants with genes for the synthesis of vaccines. This would greatly simplify vaccination because a person could simply chew on a seed or eat a banana to become immunized. It could replace expensive and less convenient conventional delivery methods for vaccines, which involve costly refrigeration and often injection by a medical officer.

Transgenic microbes are important as potential "factories" for making any type of protein. The gene for human insulin is currently working in transgenic bacteria, providing a cheap source of human insulin as an alternative to the pig or cow insulin that was used previously.

Today, private companies are being established to develop many kinds of transgenic projects. Multinational corporations such as Dow Chemical, Inco, Monsanto, and Eli Lilly have invested millions in companies such as Cetus (Berkeley) and Biogen (Switzerland) on the promise of a technology to rival microelectronics. It would seem that the only limits to this technology are those of ingenuity and imagination. Just before recent provincial elections in Ontario, the provincial government announced the establishment of a multimillion-dollar biotechnology that would, the government expected, make exciting advances in medicine, forestry, mining, agriculture, environment, and energy. Predictions include pollutant-degrading bacteria, mineral-leaching microorganisms, alcohol production from industrial waste, and nitrogen-fixing non-legumes.

human gene therapy

One special application of transgenesis is human *gene therapy*. If plants and animals can be modified by transgenesis, why not use the same approach to try to cure genetic disease? Recombinant-DNA technology has always held the promise of correcting human genetic defects, by substituting a normal gene for a defective one. Once the gene causing cystic fibrosis (the CFTR gene) was identified and its function confirmed, medical investigators turned to possible applications of this knowledge to cure cystic fibrosis patients by gene therapy. A first step in this research was to determine whether a normal CFTR gene could be transformed into isolated cells, and this was accomplished with vaccinia virus. The virus was modified with a gene for the phage T7 RNA polymerase, and the CFTR gene was then cloned in a plasmid downstream of a promoter recognized only by that polymerase. When the virus and the plasmid were introduced into cells from a cystic fibrosis patient, lacking the critical ion transport regulator, these cells acquired normal ionic regulation.

Many gene therapy centers are carrying out studies with cystic fibrosis adults, with more gene therapy tests than for any other disease. Dozens of vectors are being examined in these laboratories for efficacy, but none has all the characteristics desired in a gene therapy vector. Some gene transfers are performed by bronchoscope, a tube used to examine the lungs and, in this application, to introduce material into them; other centers prefer to treat the nose or sinuses, because they are easier to reach and mistakes are not as irrevocable.

As shown first by Ronald Crystal and his colleagues at Cornell University, adenovirus can deliver the CFTR gene to lung cells in humans, where it is expressed. Because the virus does not integrate into the human genome, repeated dosing is required; but, unfortunately, subsequent doses result in lowered transfer efficiencies and also produce inflammatory reactions. To counteract this problem, more and more of the adenoviral DNA has been removed, with the aim of producing a vector with no viral gene products except for the few needed for packaging and expressing the CFTR DNA. It is hoped that such a vector will not be recognized by the host immune system, and repeated dosing will be as efficacious as the first.

A small adeno-associated virus (AAV) is a possible vector for the CFTR gene because, unlike adenovirus, it does not cause disease. However, it does not transfer the gene so well. To improve AAV as a vector, irradiation and chemical modification of the virus is now being tested. Other laboratories are working on retroviruses carrying CFTR, because these viruses naturally integrate their genome into host cells.

The question remains whether expressing a normal CFTR protein will rid the cystic fibrosis patient of bacterial infections in the lung, which cause more than ninety percent of the morbidity and mortality. There are indications that gene therapy could do this successfully. A protein in lung secretions whose function is to kill does not work in the presence of high salt concentration, the condition characteristic of cystic fibrosis patients; but the factor becomes active and is able to kill bacteria again once the salt concentration is lowered, as it would be by an active CFTR.

Many other methods are being developed for gene therapy to treat the many known genetic diseases. Disorders associated with abnormal blood cells, for instance, might be treated by transforming those cells in tissue culture and then introducing them into a patient's bone marrow, where they find a natural home. It is all but certain that some of these methods will succeed and will gradually become part of medical practice in the next few years.

The above approaches to gene therapy are all *somatic gene therapy* – that is, they are applied to the soma (body) in the hope that enough transgenic cells can be made that they will provide enough product for normal function. Although a patient might be cured, the transgene cannot be passed on to children because the body cells modified do not include the germ line (gonads). *Germ line gene therapy* aims at genetic modification of the entire organism, including the germ line. The simplest approach (in principle) is to modify the fertilized egg by injecting a suitable transgene. This type of procedure is possible and has been successful in experimental animals such as mice. But could it be applied usefully in humans; and if it could be, should it be? This is a serious ethical issue, and some ethicists have argued that somatic gene therapy is ethical but that toying in this way with the human genome, changing the genes of our descendants, ought to be forbidden.

genomics, the study of complete genomes

Improvements in sequencing technology, and the development of robots to handle the large number of clones in a genomic library have led to the ability to sequence entire genomes. Now the complete DNA sequences of many species are available, including most of the so-called model genetic organisms such as *E. coli*; the roundworm *Caenorhabditis elegans*, used for studies of development; and the classic *Drosophila melanogaster*. In the 1990s, amid a great deal of controversy and competition among laboratories, the Human Genome Project was undertaken, eventually under sponsorship of the National Institutes of Health. In February 2001, a large group of investigators headed by J. Craig Venter, of the private laboratory Celera Genomics, announced that they had completed a good rough draft of the human genome; it was published in the 16 February 2001 issue of the journal *Science*, and another version was published in the 13 February issue of *Nature* by another large group, the International Human Genome Sequencing Consortium.

In one sense genomics has existed ever since the middle of the twentieth century when geneticists developed maps of the complete sets of chromosomes of model organisms using mapping techniques based on the frequency of recombination (Chapter 8). However, those maps showed only some of the genes, those for which mutant alleles were known, and hence they were incomplete. In contrast, complete DNA sequences contain all the genes of the organisms, and also the DNA between those genes.

Genomics is divided into structural and functional approaches. *Structural genomics* attempts to find out where the genes are along the vast tracts of DNA that constitute the chromosomes. Computers are programmed to recognize the sequences that usually characterize the beginnings and ends of genes, and so segments bearing candidate genes can be identified; these possible genes are called *open reading frames* (*ORFs*). The same type of computer programs can spot the presence of introns in the ORF sequences. Once introns are subtracted from the DNA of a candidate gene, the remaining coding sequence is translated by the computer into a protein sequence. The proposed proteins are then compared with banks of protein sequences from genes of known function that have already been sequenced. This kind of program has demonstrated a great deal of *evolutionary conservation*: that most genes have very similar counterparts in a wide range of organisms. These close similarities make sense in view of evolutionary descent from common ancestors; once a protein – and the gene that encodes it – has been shaped by evolution to be functional in one species, it should generally require little or no modification to be functional in the descendants of that species. Such evolutionary conservatism makes it relatively easy to use known genes to hunt for related genes in other organisms. After comparing a candidate gene with known genes, we can often assign it a likely function, and this can be tested in further experiments.

After the positions of the candidate genes have been deduced and charted on the complete DNA sequence, a gene map can be made. A gene map derived from the human DNA sequence is a colorful affair, with each gene colored according to its apparent function, based on comparisons with other known genes. Remember that most human genes, like eucaryotic genes generally, have many large introns; in fact, about a quarter to a third of the published sequence is estimated to consist of introns. Remarkably, only about 1.5 percent of the entire human genome (about 2.9×10^9 base pairs) appears to contain the sequences (exons) that actually encode proteins. Furthermore, this DNA appears to constitute only about 35,000–45,000 genes, quite a bit less than informed guesses had predicted. We have yet to understand the implications of this relatively small number in an organism as complicated as a human.

About two-thirds to three-quarters of the genome consists of wide spaces between genes, another contrast to bacteria. These spaces are not empty, but their contents are still something of a mystery. A large proportion of intergenic regions are occupied by *repetitive DNA*, DNA present in many copies ranging from a few hundred to many thousands of nucleotides. Some types of repetitive DNA are clustered; others are dispersed throughout the genome. Most repetitive DNA is non-functional but is derived from sequences that once probably did have function. A large class of repetitive DNA is derived from *transposons*, DNA segments that are capable of moving around the genome. Most human transposons are now inactive. These repetitive sequences have been called "junk DNA"; however, this label may simply be a reflection of our ignorance, and they may hide important unknown functions. Another class of repetitive DNA comprises inactive viral genomes that once parasitized human cells and inserted their sequences into the human chromosomes.

The number of copies of repetitive DNA varies from individual to individual. Hence they constitute useful markers of individuality. This is the basis of the forensic tests illustrated earlier.

Functional genomics is the experimental exploration of gene function at the genome-wide level. Although candidate genes are often identified by similarity to genes with known functions in other organisms, their function must be demonstrated in the genome under study. In certain model organisms, such as baker's yeast, it is possible to systematically eliminate – or *knock out* – the function of each gene in turn on a genome-wide basis. *Gene knockout* is achieved by replacing the resident, functional gene with a deleted form of the gene on a special vector. A strain whose gene has been knocked out can then be tested to evaluate the phenotype of the cells missing that function. In the ongoing program to analyze the genome of one species, baker's yeast, every one of the many thousands of genes has been knocked out.

Another approach of functional genomics is to study genome-wide patterns of transcription. This procedure acknowledges that although it is important to study the effects of single genes, most biological processes are highly complex and undoubtedly involve many genes. The developmental processes that shape an organism are of particular interest, as we indicated in Chapter 11. If transcription patterns are measured under different conditions of growth and development, the hope is that wholesale gene pathways can be pieced together.

But how is it possible to study the transcription of an entire genome? Again, new technology has shown the way. The DNA of each gene in the genome, or a partial set, is laid out in an ordered array on the surface of small pieces of glass, called *DNA chips*. These are challenged with the full set of mRNAs present in a certain cell. The DNA bound to the chips is made by one of two methods. In one method, all the mRNAs are reverse-transcribed into short gene-sized DNA molecules called *cDNAs* (complementary DNAs). In the other method, the genes (or parts of genes) are synthesized one base at a time on specific spots on the glass surface. The synthesis is carried out by robots that uncover and cover each spot on the glass in an appropriate sequence. Today, genomic DNA chips for many organisms can be purchased from the chemical companies.

To study transcription patterns, the mRNAs from a particular developmental stage of interest is labelled with a fluorescent tag and then poured over the chips. These mRNAs will stick to the DNA of their own gene on the chip and tag that spot with fluorescence. Since the position of each gene is known, the pattern of fluorescent spots fed into the computer determines which genes were being transcribed at that developmental stage.

Using techniques of this kind, geneticists are starting to identify global patterns of organization at the structural and functional levels. A special branch of biological computing called *bioinformatics* has evolved to handle the mass of information coming out of this research. The coming decades will be exciting times as the general blueprint for life is pieced together.

chapter thirteen

the geneticist as dr. frankenstein

In the public's eye today, geneticists seem to fit the image of scientists so obsessed with their work that they end up creating monsters, as Mary Shelley portrayed so powerfully in her classic novel *Frankenstein.* Like Shelley's famous doctor, geneticists are viewed as dabbling in the forbidden, creating harmful products, and even violating some natural order. Moreover, in another scene straight from Shelley's book, the citizens of the global village are rising up with their metaphorical pitchforks and storming the castle of science, as they tear up fields of genetically modified crops and protest against the "new genetics."

the regulation of recombinant-DNA research

The revolt began long before the recent explosive growth of genetic engineering. In the 1970s the scientific community and lay public became enmeshed in a protracted controversy about possible hazards of DNA technology. What exactly was the basic issue? The problem centered on the nature of recombinant DNA, also known as chimeric DNA. We have seen that DNA chimeras produced in the laboratory are new combinations of DNAs of unrelated species, combinations that could never occur in nature. Most chimeras are made for specific experimental purposes. They are controlled interventions based on a detailed understanding of the system being explored. However, sometimes chimeras are exploratory, made to "see what happens." Even though the results usually conform to what is expected, many of the most interesting results are totally unexpected. It is characteristic of science that often we do not really *know* what is going to happen until an experiment is done. So scientists cannot always be sure that cells carrying recombinant DNA will not be harmful. This uncertainty about the safety of chimeric DNA is at the heart of the issue.

It was this dilemma that molecular biologists of the 1970s faced as they contemplated experiments such as inserting genes from a cancer-causing mammalian virus into the DNA of the bacterium *E. coli*. The motives were noble: to study the function of cancer genes in a simpler system. But could this technique accidentally create bacteria that might induce cancer in people? As the scientists who had developed the methods confronted unknown potential consequences, they agonized over their social responsibility in creating potential hazards in an organism like *E. coli* that inhabits the human intestine. Consequently, in a widely publicized and unprecedented step, eleven eminent molecular biologists published a letter simultaneously in the prestigious scientific journals *Nature* and *Science* calling for a moratorium on certain kinds of experiments and caution on others. The letter asked for a complete ban on experiments with antibiotic-resistance genes, toxin genes, and cancer-causing viruses; called for an international meeting to discuss the issue; and asked the U.S. National Institutes of Health (NIH) to set up guidelines for future experiments.

For the scientific community this was a major step: in the face of unknown and speculative hazards, scientists were voluntarily refraining from doing certain kinds of experiments. The original signatories probably never imagined how much press coverage and public interest their action would arouse.

Once attention was directed to potential hazards, the public responded as though the dangers were real rather than potential. "If the experiments weren't dangerous," people often asked, "why would they have called a moratorium?" But the recombinant-DNA techniques opened up whole new areas of research, and the lure of Nobel Prizes and potential economic payoffs dazzled scientists. By the time a meeting was held to consider the ethical aspects of recombinant-DNA work, some scientists seemed more concerned with public liability than ethical responsibility; what had begun as a responsible act turned into retrenchment against public involvement in the work. Although the original highly publicized controversy has now faded, it is instructive to review some of its highlights.

On 24–27 February 1975 a group of internationally recognized molecular biologists assembled to discuss the issue at Asilomar, near Monterey, California, the setting for John Steinbeck's novel *Cannery Row.* Some attendees downplayed the ethical issues and urged getting on with the important research; others feared legislated control or were concerned with legal liability in the event of a health problem. Many eminent people clearly felt the whole exercise was a waste of time. The conference ended by giving cautious approval to go ahead with experiments, replacing the moratorium with safety guidelines for different experiments based on their estimated levels of risk. Two precautions were suggested:

1. increasing levels of physical containment for experiments posing increasing levels of potential risk;
2. use of genetically enfeebled strains of microorganisms that could survive only under special laboratory conditions.

The overriding difficulty was then, and remains today, that of estimating risk levels for experiments presenting entirely novel situations.

The NIH, which funds a large share of basic research in biology, took the initiative in establishing guidelines and in holding discussions about them in public. To define rules for research, it established a Recombinant DNA Committee consisting of scientists, experts from various fields, and representatives from private companies that were being established to apply the new technology. The arguments still revolved around possible dangers. Though one might have expected industrial representatives to oppose the whole process and seek minimal restrictions, they actually took a responsible stance. Recognizing the uncertainties, and fully aware of their legal and financial liabilities in the event of damage caused by their work, they were eager to have guidelines that would prevent disasters. Meanwhile, other scientists were clearly chafing under the public scrutiny and restrictions, which they countered by pointing out the possibility of recombinant-DNA research solving global problems such as famine and infectious disease.

Finally, on 23 June 1976, Donald Frederikson, director of the NIH, announced formal guidelines for recombinant DNA that would apply to all recipients of NIH grants. The guidelines defined four levels of *physical containment* to which experiments are assigned on the basis of their estimated risks. P1 facilities are for harmless experiments requiring only standard microbiological techniques. At each higher level, precautions become progressively more stringent, so that the P4 level – which recalls the facility depicted in the film *The Andromeda Strain* – is so rigid that no laboratory meeting the criteria existed until 1978. In addition, three types of *E. coli* were defined for *biological containment*. Standard lab strains are designated EK1. EK2 includes specially mutated strains that have less than 1×10^{-8} chance of surviving outside a lab. EK3 requires strains like those used for EK2 that have failed to survive in tests on animals, plants, or outside the lab. To produce an enfeebled strain of *E. coli* into which recombinant DNA could be inserted, Roy Curtiss – a member of the NIH committee – engineered a strain chi-1776 (in honor of the American bicentenary) containing fifteen separate blocks to its normal reproduction.

Meanwhile, the role that an educated public can play in guiding and regulating possibly dangerous research became clear through an episode in Cambridge, Massachusetts. On the day the guidelines were announced, an extraordinary public meeting was held by Alfred Velucci, the mayor of Cambridge, to consider the Harvard scientist Mark Ptashne's application to build a special lab to transfer genes of the animal virus SV40 into *E. coli.* One of us (D.S.) attended this meeting. As the home of both MIT and Harvard, Cambridge would inevitably have numerous labs engaged in gene-splicing experiments, but all new building on the two campuses has to be approved by the city council, so Ptashne's application to build P3-containment facilities stimulated the open council meeting.

For two and a half hours, before dozens of television, radio, and newspaper reporters, and several hundred citizens, advocates and opponents testified before the council. The scientists seeking approval generally felt that the work was necessary for research related to cancer, while the danger of creating a dangerous bacterium was "extraordinarily small." Arrayed against them was a long list of scientists and lay people who argued that even in the highest containment labs hundreds of accidents already had occurred, and that if a dangerous organism were produced, there could be no way to stop it. Pointed questions from the mayor and councilors showed they were listening and had done their homework. Eventually the mayor called for a two-year moratorium on all recombinant-DNA work going on in Cambridge, but the city council instead established a Citizen's Experimentation Review Board (CERB) composed of eight non-scientists. The four men and four women included a physician, a philosopher of science, a fuel oil distributor, a structural engineer, a clerk, a nurse, a social worker, and a housewife. Members of the Board educated themselves on the esoteric details of molecular biology before coming to a unanimous decision in January 1977 to allow Ptashne to build his laboratory. This precedent-setting committee and decision broke ground for other local public involvement in debates on recombinant DNA. Its take-home lesson was that ordinary citizens could responsibly involve themselves in serious issues of public policy regarding science and could reach sensible conclusions, neither inhibiting valuable research nor endangering the public.

In Britain the recombinant-DNA issue was resolved differently. A government document established a Genetic Manipulation Advisory Group (GMAG) made up of politicians, scientists, and union representatives. All proposed experiments on recombinant DNA in Britain must be assessed for approval or rejection by GMAG in light of current knowledge. The major differences from the American experience are the absence of public meetings, the inclusion of private, government, and university sectors under GMAG, and assessments that change as more information becomes available. At least eight other European countries have now set up committees to determine conditions for gene splicing. In Canada, the Medical Research Council follows the NIH conditions, but includes mammalian viruses and cell cultures for consideration.

genetically modified organisms

The issues of public involvement and imposing limits to scientific inquiry are still far from settled. The debate serves as a vivid illustration of the impact of science on society, and it has taken on new dimensions as the use of recombinant-DNA methods has become routine in labs around the world. The issues now are not simply about adding a few new genes to bacteria in the lab but about the creation of *genetically modified organisms* (*GMOs*), transgenic animals and plants designed and created for specific research and commercial uses. Many transgenics are still under test, but some have been released into the market. Transgenic technology has engendered heartfelt opinions, as high economic stakes clash with issues such as human health and environmental protection. In the end, the fundamental issue, just like the issue in the public debate in Cambridge, is whether inserting alien DNA into an organism can have unexpected and deleterious consequences. The decade or so of experience with GMOs has indeed shown that the technology is not without hazard. For example, the use of human insulin made by transgenic bacteria has resulted in some adverse reactions to the product. However, so far there have been no doomsday scenarios. The NIH has revised its guidelines, giving scientists considerable freedom to carry out experiments using recombinant DNA, although the most obviously dangerous applications are still well controlled. But the issues surrounding GMOs remain, and it is to these that we now turn.

Two of the first GMOs to be developed were ice nucleation bacteria and "Flavrsaver" tomatoes, both introduced in the 1970s. The bacteria had been modified so they could be sprayed over crops and act as nucleation sites for ice crystals; this makes crops partially resistant to frost damage and lengthens the growing seasons available to farmers. The tomatoes were engineered for delayed maturation, to increase their shelf life and thus reduce wastage from softening. Both products were eventually banished after public outcries.

Most geneticists treated these cases as storms in teacups, and a period of relative calm followed. However, two incidents alerted geneticists that some elements in society were viewing them very differently. First, in a widely publicized 1993 case, the so-called Unabomber (later identified as Theodore Kaczynski) sent a letter bomb to Charles Epstein, a prominent American geneticist, to protest the "new genetics." Then in 1996 photographs appeared in papers around the world showing members of the Greenpeace organization engaged in a large, well-orchestrated protest in Liverpool harbor involving flotillas of Zodiac boats, huge banners, and projection of slogans on the sides of ships. Their target was freighters delivering transgenic soybeans from North America to Europe. Their banners screamed, "Stop genetic pollution." For

Vancouver geneticists (such as A.G. and D.S.), this was a double shock, since Greenpeace was born in that city; its members were well known and respected for their protests against whaling and nuclear testing. What a shock to learn that Greenpeace had expanded its targets from evil whalers and nuclear powers to evil geneticists!

In the last few years the pace of protest has quickened and it now embraces most of the developed world. Huge anti-GMO parades have become a common sight, and radical groups have opposed GMOs with civil disobedience by destroying transgenic plants and even whole research facilities. The front cover of a recent *Economist* summed it all up with a drawing of a "Frankenfood" potato exclaiming, "Who's afraid of GM foods?" Though scientists originally envisioned GM foods helping to feed the poor and hungry masses of the world, so far this has not happened. We examine this issue in the coming sections.

technology in context

One side of the issue is public education about science. Real estate brokers joke that the three key elements for making house sales are location, location, and location; similarly, in science education the three keys to learning are context, context, and context. Without context, the new discoveries of science are like floating balloons of knowledge, untethered to the rest of human activity. Much of the controversy and suspicion surrounding the new genetics can be attributed to a lack of suitable context for genetic advances, and we must consider the new genetic technology in the light of technology in general.

All technology has positive and negative attributes. Most people would agree that the Industrial Revolution of the nineteenth century, based on scientific advances in physics and chemistry, has been an overall success in raising the standards of living in industrialized countries. However, the downsides of this technology are immense. One product of the Industrial Revolution, for example, is the internal combustion engine that powers vehicles and numerous other machines. The automobile is beloved by most for the freedom it provides, yet it is a massive source of ecological and health problems through air pollution, mining for metals and oil, and deforestation and loss of biodiversity owing to highway construction, urban sprawl, and the planting of rubber trees for tires. In Canada alone, about 16,000 deaths each year can be directly connected to atmospheric pollution from traffic. To these must be added the many thousands who die from traffic accidents. These deaths are real, not imaginary. Although these are tragedies to those concerned and their families and friends, society tolerates them as an unavoidable by-product of the internal combustion engine.

Likewise, the chemical industry spawns vast amounts of pollution. The chemical revolution of the previous century ("Better things for better living, through chemistry") has given us plastics, dyes, and many other useful materials. But it has also given us a food chain poisoned by insecticides, the hole in the ozone layer, chemical pollution of most natural bodies of water on the planet, radioactive contamination, and thousands of festering toxic lagoons around the world. Chemical pollution is responsible for not only human ill health and death but for the death of countless plants and animals in natural ecosystems. Genetically modified organisms must be placed in this context. Undoubtedly, modern biotechnology will have its negative aspects – social, ecological, and medical – even though to date there are very few reports of ill health stemming from the use of GMOs. The potential ills of genetic technology must be weighed against the potential good, and genetic technology must be stacked up alongside other technologies.

The current debate over biotechnology should be informed by the past, and it may have an important role in shaping the future of all technology. Having now seen all the negative consequences of past technologies, humanity should be able to take the obvious lesson: be very careful! This is the *precautionary principle*: given a choice of policies and limited ability to predict the consequences of our actions, choose the policy most likely to avert a disaster and allow the greatest flexibility. In general, that means *not* undertaking a project until it has been thoroughly examined. This principle is based on the historical precedent of having dealt with unprecedented technologies, and the ability to transfer genes from one species to another is without precedent. When DDT was found to kill insects, it was hailed as a revolutionary method of pest control. Its discoverer, Paul Mueller, received a Nobel Prize in 1948. Although geneticists knew resistant mutants would be selected, and ecologists knew insect pests are only a small proportion of all insects, a broad-spectrum insecticide seemed like an acceptable pest control measure. But no one knew that DDT and many other chemicals would be concentrated up the food chain because they collect in fatty tissue, so each animal retains all the chemicals in the tissues of its food. In this process, called *biomagnification*, these chemicals become concentrated hundreds of thousands of times and reach critical levels in the shell glands of birds and the breasts of women. This phenomenon was only discovered when raptors began to disappear. No amount of caution could have avoided this unknown hazard. Nor could we have known the dangers of chlorofluorocarbons (CFCs), which were hailed as miracles of chemistry when first synthesized. Chemically inert, CFCs made perfect carriers for chemicals dispensed in spray cans. No one knew that CFCs would persist in the upper atmosphere and there would produce free chlorine radicals that scavenge ozone. The nature of revolutionary technology is that our knowledge base is too limited to anticipate the long-term consequences.

the arguments against producing gmos

Several kinds of arguments have been used to build a case against work on genetically modified organisms. We review these here so readers will have a basis for forming their own opinions. We do not conclude by taking sides, but clearly some of the arguments stand up to scrutiny better than others.

the unpredictability of genetic effects

Supporters of transgenic research argue that transgenic DNA is just DNA and genetics knows a lot about DNA, so why worry? However, our knowledge of how DNA behaves in heredity is based largely on manipulations *within* a species. Transgenic work moves one species's DNA into a totally different species, and it may be dangerous to assume that behavior across species is the same as within a species. To put this in perspective, consider the lesson of Chapter 11: that a genome functions in a wonderfully orchestrated sequence through development. A gene is not expressed in isolation. Rather, it functions as one element of the total genome. Conventional breeding experiments in both plants and animals have shown that changes in a few genes can have profound effects on a developmental program. Most of the mutations that have been manipulated in conventional breeding are *pleiotropic* – that is, they affect several genetic networks and have multiple effects. They can produce dramatic changes in appearance but in no sense do they create monsters. For example, the food plants cabbage, cauliflower, broccoli, kale, kohlrabi, and brussels sprouts are all strikingly different in appearance, yet they were produced by mutations in one wild species of plant, *Brassica oleracea*. Likewise, the many breeds of domestic dogs were produced by mutation and selection in the original wild species of dog, affecting very few of the thousands of dog genes. Although single mutations can produce quite distinctive organisms, they rarely produce monsters.

Now, a piece of alien DNA inserted into a cell finds itself in a new context. We cannot be certain how it will function, but we can make some reasonable guesses, based on our current knowledge of cell biology. It's somewhat like putting Mick Jagger of the Rolling Stones into the New York Philharmonic Orchestra. A good conductor (read: a geneticist) will have a pretty good idea of how the rock musician's guitar and style could fit into that of the orchestra, but until the performance actually takes place the conductor cannot be sure what the sound will be like. The dilemma for transgenic work is that the potential benefits of DNA manipulations can be planned and expected according to the current understanding of the genes concerned, but the possible negative side effects are more difficult to anticipate.

In years of genetic research, many genes have already been transferred into quite different genomes without creating monsters. The first transgenic mammal ever made was a mouse that carried a rat's growth hormone gene. As predicted, the mouse grew much larger than its brothers and sisters (and looked like a rat), but suffered no obvious health problems associated with the transgene. Today geneticists routinely use transgenic organisms in their experiments. Genes from a firefly have been inserted in a plant to make it glow in the dark, and genes from a jellyfish can be inserted into mice, also giving them a glowing appearance. Genes from bacteria have been inserted into fruit flies and plants, making them turn blue in the tissues where this transgene is expressed. These genotypes are created in a highly controlled and deliberate way; the transgenes are often extensively tailored to suit the requirements of the recipient genome. Plant breeders have also extensively hybridized different plants, making combinations that probably would not have occurred in nature. The crop species triticale is a human-made hybrid of wheat and rye, which belong to distinct plant genera. This unnatural fusion provided ample opportunity for adverse genetic interactions, since two very different genomes were brought together, yet triticale is a rugged crop that provides considerable health benefits. Nor have these mutant types overrun natural ecosystems. Thus the accumulated experience of research provides an important context for new applications of genetic technology and must be considered in each new potential application of genetics.

Finally, sequencing the genomes of humans and other organisms has shown that our genomes contain numerous genes from other organisms (such as bacteria); they must have been acquired by unknown mechanisms, perhaps transfer by viruses. This is called *horizontal transmission*, in contrast to ordinary *vertical transmission* of genes from parents to offspring. Therefore in one sense we are all naturally transgenic.

negative health effects from gm foods

The organisms we eat are already complex mixtures of organic compounds, which are both benefits and hazards to health. Research by nutritionists and physiologists regularly reveals that certain foods have previously unknown benefits. The tannins in red wines and the carotenes in tomatoes have salubrious effects on cardiovascular fitness and preventing cancers. On the other hand, many compounds in common foods are known to have negative effects on health. Some are even carcinogens, such as compounds in black pepper and the brown surface of roasted and grilled meat. Also, many people have potentially deadly food allergies. The general fear is that adding foreign genes to a food plant may produce materials with unknown effects, and especially that foreign proteins may be new sources of allergies. In 2001, an independent scientific panel reported that there is a "medium probability" that a protein (called Cry9C) in the GM StarLink corn is a human allergen. U.S. agencies, including the

Department of Agriculture, continue to try to divert StarLink corn from the human food supply, while maintaining that the process of wet-milling corn removes virtually all of the StarLink protein from products made for human food.

Some crops, such as corn, have been transgenically modified with genes from the bacterium *Bacillus thuringensis* (known as Bt), which provide a toxin that kills certain insects that can grow on these plants. The possible effects on humans of foods with the Bt proteins are unknown, yet these foods are now on the market. A report issued by the esteemed scientists of the U.S. National Academy of Sciences finds no scientific evidence of harmful effects of GM foods, although there has not yet been time to judge possible long-term effects. We note in passing that there is a certain irony in a recent move by Canada's and America's biggest potato growers, who supply the multinational fast-food chains: in spite of the documented evidence of real and massive health risks from the fat and cholesterol in the fast-food products, the growers have decided to withdraw all transgenic potato stocks on the grounds of possible "negative effects on health."

potential ecological damage

Crops grow in open fields, where GM plants have many opportunities to hybridize with other plants and where vectors such as viruses could transfer some of their genes to other plants. On the other hand, a large proportion of the world's arable land is currently planted with specially bred crop species not native to the region, so there is already ample opportunity for this type of gene dispersion. Invasive species all over the planet are already causing massive ecological crises.

Plants containing genes such as those for Bt toxins will act as strong selective agents on insects; since sensitive insects will be killed massively by the huge fields of such crops, resistant insects will soon be selected. This will have unknown ecological consequences. One study reported that pollen containing Bt toxin was killing monarch butterflies, but further investigation showed that this is not a problem in the wild. The threat of such a toxin pales before the habitat destruction that threatens the migrations and hence the survival of this same species; but this simply means that we must maintain a broad viewpoint and try to eradicate all ecological threats from human activity together. Note that insect resistance is not unique to Bt crops: fields sprayed with insecticides also provide environments that select strongly for resistance.

increased power and profit for transnational corporations

Transgenic biotechnology is expensive, and corporate research in this field has relied heavily on funding by speculators who demand a high return on their investments. Transgenics are therefore generally developed to increase profits, rather than for the benefit of the poor. So it is no wonder that many people perceive both health and environment to be under threat from "genetic pollution" that is "unpredictable, uncontrollable, unnecessary and unwanted" (Greenpeace).

In this connection it is important to recall the distinction we made in Chapter 1 between the relatively few academic scientists who do fundamental research and the much larger numbers now employed by government and industry. The concerns of academics for advancing knowledge and improving society often contrast with the desire of industries, and the scientists they employ, to profit from the new technology. It is not that academics are the "good guys" and companies are the "bad guys." However, it is generally true that academics make progress cautiously and are slow to jump to conclusions, whereas companies are under constant pressure to satisfy their investors by showing profits every year. This fundamental tension is an important factor in the GMO dilemma. It is partly the tension between technology, which is ethically neutral, and the applications of technology, which may be ethically dubious.

Some corporations have sought worldwide dominance of an industry. For instance, Marc Lappé and Britt Bailey have reported (for Alternative Radio on U.S. public radio stations) on their experience with the soybean industry and efforts by the Monsanto Corporation[1] to control it. Monsanto has introduced genes into soybeans that make the plants resistant to the herbicide glyphosate, which it sells under the brand name Roundup. Monsanto's goal is to control soybean production worldwide so all farmers will use their Roundup Ready strain and will, therefore, use Roundup at high levels on their crops; by 1999, it already controlled two-thirds of the

American market. Monsanto owns the seeds, and farmers are forced to purchase new seeds every year rather than keeping back some seeds for replanting the next year, as is the traditional practice. Anyone who tries to save seed to grow a new crop can be sued. Lappé and Bailey encountered fierce resistance from Monsanto in their investigations, and they uncovered considerable evidence of deception by the corporation. Meanwhile, soybeans – a major source of the world's protein – are being produced increasingly with proteins whose effects are unknown. They apparently also have altered levels of phytoestrogens, the plant analogues of mammalian hormones, and no one knows the health consequences of such a change. Furthermore, people are being forced to consume more and more soybeans sprayed with Roundup, with additional unknown health consequences. Again, however, note that the problem is not unique to GM foods: herbicides were used massively for much of the twentieth century. The impact of such exposure on human health is not known, but regulating herbicide use for reasons of health is an issue independent of the issue of GM crops.

The actions of large corporations paint industry as villain. Supporters will argue that industry is just using genetics as it has used other technologies, to achieve a competitive edge and maximize profit. Agricultural companies, for instance, have genetically engineered desirable crops by introducing genes called *terminators*, which make their seeds infertile – again forcing farmers to purchase seeds anew each year. One of the original intentions was commendable: to prevent genes from these plants escaping into other populations. This also amounts to a type of patenting. On the one hand, the potential injustices of patenting genetic information are not different in principle from other types of patenting, and the suffering of farmers associated with terminator corn stocks is no different from the suffering caused by any other type of protection of marketable ideas. On the other hand, biotechnology gives industries far greater potential for control (and harm) than they have ever had before, and this raises serious economic and ethical questions. Perhaps affluent American agriculture can afford to use terminator-bearing plants, passing the extra cost on to the affluent American public; but how will the cost be borne by billions of people in developing countries already living on the edge of starvation?

If ethics is to mean anything at all in modern society, can we allow any industry to enrich itself at the expense of the public as much as unregulated biotechnology might allow? Perhaps the debate over biotechnology will force humanity to come to grips with the broader issue of what science and technology are for. Are they to be used for the often stated ideal, the benefit of all people, or only to enrich a few? Is society reaching a breaking point, where it will no longer tolerate industrial abuses of human rights, whether from biotechnology or any other technology? Perhaps the current debate will focus humanity's need to deal with the general consequences of technology, as we have outlined them here.

Lastly, the issue of bottom-line profits is a slippery one. It is easy to point the finger at corporations, but the invested savings of many ordinary people benefit from high interest rates based on these profits. The context for these issues is a complex economic system of our own making. The answer of the neoclassical economists who largely guide our society is to have faith in Adam Smith's "invisible hand of the marketplace." They will insist that if technologists and entrepreneurs are allowed to freely develop technologies that will benefit them economically, the whole society will benefit. Our experience over the past two centuries should make us, at the very least, extremely skeptical of this stance.[2] Perhaps, as the wise cartoon character Pogo maintained, "We have met the enemy and he is us."

dna technology is highly unnatural (scientists "playing god")

Transgenic modification has been singled out as "unnatural." Indeed it is, but so is all technology, including the selective breeding that humans have carried out for thousands of years. Sheep, cattle, pigs, and poultry have all been modified to suit their nutritive or other properties to our preferences and perceived needs. The food we eat and the clothes we wear come from plants and animals genetically modified by traditional breeding. Transgenic corn is no more unnatural than the many mutant forms of plants and animals used in farms throughout the world; but whereas one delights, the other horrifies. Gene therapy is in principle no more unnatural than therapy by drugs or by surgery.

cloning as an ethical target

The cloning of mammals, although not a case of transgenesis, raises some of the same ethical issues. Although animals such as toads had already been cloned, the issue broke into public consciousness with the much heralded birth of Dolly the sheep, a cloned copy of a sheep who was not her mother. The subsequent reports of cloning in other agricultural animals has fuelled the case that geneticists have emulated Dr. Frankenstein or even played God by interfering with the natural order of reproduction. These cases naturally raised the specter of human cloning, which seems to be universally loathed.

The word "cloning" itself has caused some confusion. "Clone" originally meant a group of genetically identical organisms formed asexually, such as all the bacterial cells formed by repeated division from one original cell. (Note that "clone" is sometimes used to mean an individual member of the clone. Although technically incorrect, this usage has become entrenched and is now accepted.) *DNA cloning*, as described in Chapter 12, is the method used in transgenesis: inserting a specific piece of DNA in a vector and replicating the vector inside cells to generate copies of the DNA. However, when biologists speak of cloning an organism such as a sheep or a human, they mean quite a different process. In this procedure, the nucleus of a fertilized egg is removed, discarded, and replaced with a nucleus from a somatic cell of another individual. The nucleus that was discarded contains a mixture of chromosomes from the mother and father; but the somatic-cell nucleus contains only the chromosomes of one individual, the donor, so the resulting individual should be genetically identical to the donor. The procedure was first developed in frogs and toads to test ideas about the genetic regulation of embryonic development. The formation of identical twins (triplets, quadruplets, and the like) is a kind of natural cloning in humans. Identical siblings result from an accidental division of a zygote to make two or more cells, which then develop into genetically identical people.

Dolly the sheep was made with the nucleus from a mammary gland cell of the donor female. (Dolly's name was inspired by the country-and-western singer Dolly Parton, who is famous for her mammary glands.) This cell was allowed to divide until it came to a stage of the cell cycle at which it could act as the nucleus of a zygote. The same type of technology has been applied to clone other mammals with success. In 2001 some laboratories reported the successful cloning of humans, but the resulting embryos were destroyed before they could develop very far. (This work was never subjected to the rigors of peer-reviewed publication.)

What are the objections to cloning humans? One argument centers on an early report of the ill health of Dolly. However, this seems to have been exaggerated, and even if it is true, improvements in the technology could probably correct tendencies toward defects. The major objections are all ethical. First, the nucleus of a normal zygote must be destroyed, and this zygote had the potential to become a certain human being. Second, Dolly was only created after many trials, some of which produced abnormal individuals, which were discarded. Third, cloning might be abused to produce armies of workers, soldiers, or any type of individual at the whim of an unscrupulous government or company. Fourth, in cloning a human, scientists are "playing God" by interfering with the normal mode of human reproduction. Fifth, there is a fear of general prejudice against cloned people, who might be branded inferior by virtue of their origin. Lastly, there is the fear that cloning would be used for a type of eugenics, creating a specific set of desired genotypes, much as the Nazis tried to create a "master race" in the 1930s and 1940s. (We discuss eugenics in Chapter 15.)

Although again it is important to distinguish a technology from the unethical application of that technology, these objections must be taken seriously. The specter of a cloned society as portrayed in Aldous Huxley's novel *Brave New World* is abhorrent to most people. The common reaction against that odious image makes it seem unlikely that cloning could ever become widespread. Much of our human condition revolves around love and respect for our life partners, and having a child together is a manifestation of this love. Hence cloning humans seems likely to be relegated to the domain of rich eccentrics. Still, we have seen that once it becomes technically possible to do something, some people become obsessed by doing it. The ability to clone people and to otherwise manipulate our genes could be a force that will drive humanity toward Huxley's dystopia.

The most serious proposals for using human cloning have focused not on cloning whole people but on *therapeutic* cloning. Cell cloning can be used as a source of cells for corrective gene therapy. It might also be possible to use cloning to produce replacement organs, such as a new heart or a kidney. Such cell or organ replacements would have the considerable advantage of being genetically identical to the cells of the individual being treated, thereby bypassing problems of immune rejection. This approach relies on *stem cells*, embryonic cells that are uncommitted to any one developmental fate. Stem cells can be obtained from certain sites in a person's body, but the more controversial procedure is deriving stem cells from disrupted embryos, which is currently more effective. To obtain embryonic stem cells, a donor nucleus is injected into an enucleated fertilized egg. When this cell divides it would normally produce a new individual who would be a copy ("clone") of the donor. However, the cells of the embryo can be dispersed, grown, and used as stem cells for replacement. Stem cell research has been severely limited by legislation in the U.S., but in Europe the regulations are less severe.

Genetic technology applied to humans has provided some interesting yet controversial genetic dilemmas. Some interventions – amniocentesis, *in vitro* fertilization, somatic cell gene therapy, and even surrogate motherhood – have become largely acceptable. However, procedures such as cloning are still highly controversial, and we have yet to see the full social impact of procedures such as somatic-cell gene therapy.

the responsibility of scientists

Current genetic technology has the potential for several kinds of harm, to which society must remain alert. It is not our purpose to offer an apologia for genetic technology or to condemn it. Our point is that we must keep the issue in perspective in the context of technology in general. Our experience with previous technological catastrophes should have taught us the importance of going slowly, testing the impact of new technologies, and being guided by the precautionary principle. However, *all* technologies must be tested and assessed, including those that are currently far more menacing than biotechnology. New potent chemicals, for instance, are being developed and used faster than they can be tested. It is important for citizens to gain control over technologies and to educate themselves so they can rationally consider the value of a proposed innovation and the possible outcomes of our actions. Unfortunately, history shows us that this is not easy for humans to do.

What is the responsibility of the scientific community to the taxpayers who not only subsidize most research but also benefit or suffer from its effects? How far do scientists recognize that responsibility? Although supporters of research into DNA technology are quick to justify the work by its potential remedies for cancer, hereditary disease, pollution, famine, and overpopulation, many are also quick to deny the public the right to inspect or to set priorities in research. The Cambridge debate focused on criticism of the NIH guidelines. As Cambridge city councilor David Clem reiterated, "The principle should be that an organization that promulgates research should not also set its own rules." Agreeing with this principle, many people charge that the scientific community, having a vested interest in continuing research for recognition, promotion, and prizes, cannot be expected to keep a watchful and critical eye on work potentially harmful to society. Some scientists themselves feel that a hierarchy of influence within the scientific community exerts enormous pressure on critics of establishment policy.

Scientists who are excited by and committed to research will naturally be impatient to carry on without roadblocks. The potential rewards of transgenic research are enormous. Scientific acclaim already attends virtually every innovation, and Nobel Prizes have been going regularly to the pioneers in recombinant-DNA work and those who are now using these methods. The opportunities seem limitless. Hence the frustration of scientists with public debate; yet their impatience illustrates the difficulties of remaining objective when one's own research is at stake.

Today's young scientists, who will be conducting most of the research, ought to be leaders in keeping that work socially responsible and ethical. Unfortunately, they have been forced to specialize early in their careers to keep abreast of the rapid accumulation of knowledge, and it is important for them to develop a much broader perspective. Scientists may be paying the price of a kind of myopia and tunnel vision, which is increasing just when society is most affected by the products of science.

Geneticists who have traditionally taken great pride in the purity of their research have been dismayed to see how much financial returns are driving current research. Many leaders in the new genetics are associated with private research companies, which raises the issue that scientists who have been supported by public funding for years are now extending that publicy funded research into areas that will be profitable to them personally. Should they be prevented from doing so? And if they were, how far would it inhibit research?

Some eminent scientists have sharply criticized the hubris of molecular technology and have condemned the headlong rush to exploit the techniques. In spite of the stated hopes of this research for cures for disease, pollution, and other ills, the history of science informs us that hopes and intentions often bear no resemblance to what actually happens. As Philip Abelson, the editor of *Science*, wrote before the advent of DNA technologies:

> Geneticists will have high ideals for the application of their research. In practice, power to apply that knowledge, as was the case with nuclear energy, will come to rest in other hands.

GENOMICS AND HEALTH CARE

A health issue quite separate from that of GMOs has arisen from modern genomics. As more and more human genes are identified and as tests are developed for DNA markers showing predisposition to disease and death, questions arise about how each person's genomic information might be used. In a health care system like that of E.U. countries, in which everyone is cared for according to their needs, a detailed genetic profile could be valuable. Knowing that one is predisposed for a specific disorder, one can have routine tests to monitor for that disorder; and more therapies are being developed that can alleviate or prevent certain disorders if they are identified early. However, in a health care system like that of the U.S., where everyone must buy private insurance from companies whose primary concern is profit, there is the real possibility of discriminating against people on the basis of their genomes. This is not fundamentally a problem of genetics; it is a social and political issue about how genetic information is to be used.

chapter fourteen

the fountain of change: mutation

Throughout history, the sudden appearance of someone with a very unusual appearance has excited awe, reverence, or fear. The mythologies of many cultures document fantastic creatures resulting from the whimsical or purposeful acts of gods or the chance conjunction of environmental conditions. In the real world these creatures were either developmental anomalies or people we would now call *mutants*, who had acquired a specific hereditary defect, or *mutation*. (The event that produces a mutation is also called mutation.) We traditionally limit the term "mutation" to relatively small changes in a single nucleotide pair or a short stretch of nucleotides, in the DNA; larger damages that create visible changes in chromosome structure are called *chromosomal aberrations*. As we have come to understand the mechanism of mutation, we have acquired some control over it; but at the same time there have been enormous increases in environmental factors that produce mutations, especially radiation and chemical contaminants, whose real impact we are just beginning to fathom.

Lurid science fiction movies and stories promote the general feeling that all mutations are evil. It is certainly true that a typical accidental change will create a defect rather than an improvement; a complex organism, operating on the instructions of thousands of genes, is not likely to be improved by some random fiddling with the structure of those genes. But mutations are the basis for all the variability on which natural selection acts. Without mutation in the broadest sense, there could be no evolution. Many mutations have little or no effect on the organisms in which they occur; but organisms live in constantly changing conditions, and any mutation could confer an important advantage in other circumstances. Indeed, all natural populations carry several alleles at many loci, and some populations survive only because one allele is good in one circumstance and an alternative allele in another. (One classic study of land snails in Britain shows that several alleles that produce different shell colors and patterns are valuable because they camouflage snails in different habitats as the seasons change.) Many human proteins with several allelic forms are known. So in natural populations there is no such thing as a normal or wild-type allele, and the terminology is certainly useless for humans; blond hair may be the most prevalent phenotype in Sweden, but not in Italy or Japan. With laboratory organisms, we define a wild-type allele arbitrarily.

Mutations can occur sporadically and unpredictably at any time and in any cell, including both the germinal (gonadal) and somatic (body) cells of a plant or animal. Germinal mutations can be passed on to the progeny of the next generation. Somatic mutations are only passed on to the daughter cells of the original mutated cell, and they are major sources of cancer.

mutation rates

Mutations are always occurring naturally, spontaneously, and without any obvious cause. We can never predict when and where any particular mutation will occur, so we must treat mutation statistically. The *mutation rate* is defined as the *probability per cell* that a mutation will occur in each generation. We can measure spontaneous mutation rates, and any agent, such as radiation or a chemical, that increases this rate is called a *mutagen*. Since mutations in somatic cells are causes of cancers, mutagenic agents are also potential carcinogens.

Where it is relatively easy to determine mutation rates, in bacteria and other microorganisms, they fall in the range of about 10^{-6} to 10^{-8}; this means that a particular gene will mutate in only one cell per million to one cell per hundred million with each generation. Getting reliable measurements of mutation rates in complex diploid organisms is much harder. First, many phenotypes can be produced by mutations in a number of genes, especially in a complex organism. Second, in an organism that goes through an extended development, as humans do, an abnormal phenotype might owe to some interference in development, not to a mutation. For instance, mutations can produce deformed or missing limbs, but a pregnant woman can produce similar deformities in a fetus by using certain drugs. Third, if an organism were homozygous for a wild-type allele, it would take two essentially identical mutations to produce a homozygous recessive phenotype.

A straightforward way to study mutation in plants and animals is to look for mutation from a dominant to a recessive allele in crosses between an *AA* and an *aa* type. All the progeny should be heterozygotes and show the dominant phenotype; one with the recessive phenotype must have resulted from a mutation in the *AA* parent. The problem is that, since mutation rates are so low, one must examine enormous numbers of animals to get reliable data. This is virtually impossible with humans and difficult even with experimental animals like mice. The geneticist Lewis J. Stadler studied mutation in maize, using several loci that affect kernel phenotypes so he could see a recessive phenotype just by looking at the seeds in each corn cob. By examining millions of seeds, he found that most spontaneous mutations occur at quite low rates, typically about 10^{-6} per gamete. A few exceptional gene loci have mutation rates around 10^{-4}, a much higher rate than the measured rates for bacteria.

William Russell and his collaborators studied mutation in mice in one of the largest mouse laboratories in the world, at Oak Ridge, Tennessee. As in Stadler's experiments, they looked for rare homozygous recessive traits resulting from a mutation in the cross *AA BB CC DD PP SS SeSe* × *aa bb cc dd pp ss sese*, where all the symbols refer to genes that affect mouse fur phenotypes. Out of 288,616 mice, they recovered only seventeen mutations at all seven loci, an average of 0.006 percent. Their estimate of the spontaneous mutation rate per locus is from 1×10^{-7} to 3×10^{-7} per gamete – again, low and comparable to bacterial rates.

mutation in humans

Mutation rates in humans have been measured with pedigrees in which an unambiguous dominant trait appears. A defect that appears suddenly in one person and is transmitted to that person's children alone must have resulted from a mutation. A well-known dominant phenotype in humans is achondroplastic dwarfism, which was depicted in ancient drawings and statues; it is useful for measuring mutation rates because the mutant allele has a very high *penetrance*, which means that all who have it show a mutant phenotype. Among 94,075 children born in one hospital in Copenhagen, eight dwarfs were recorded from normal parents. This is about one in twelve thousand births. Since each child represents the combination of two genes, either of which could have mutated, the mutation rate per allele is one in twenty-four thousand or about 4×10^{-5}. In a study of retinoblastoma, in which tumors develop in the retina of the eye, forty-nine affected children born of normal parents were observed in 1,054,985 births; this makes the estimated mutation rate for retinoblastoma about 1.8 per hundred thousand gametes. The mutation rates for these and a few other traits seem high when compared with those from other organisms, but this may mean that some of these traits can arise through mutations in several genes, so each number is the sum of rates for those genes.

It is easier to infer when a sex-linked recessive mutation, rather than an autosomal recessive, arose. Simply by looking at males in both lines of a woman's ancestral pedigree, we can determine the genotype of her two X chromosomes; then if she has a son with an X-linked trait, we can tell if it is the result of a mutation. The most famous case of an apparently new sex-linked mutation was the hemophilia that afflicted several royal European lines. It must have arisen either in one of the gametes that produced Queen Victoria (1819–1901) or in one of her cells during development. Her descendants carried this mutation into the royal lines of Russia and Spain. Tsarevich Alexis, the only son of the last Tsar of Russia, Nicholas II, was a hemophiliac. In efforts to cure his disease, his mother, Tsarina Alexandra, turned to faith healers and eventually ended up in the grip of Rasputin.

radiation

Spontaneous mutations are rare events. Mutation rates are increased by mutagens, and among the most powerful mutagens are some types of radiation. In 1927, Herman J. Muller, using *Drosophila*, and L.J. Stadler, using maize, independently discovered that exposing these organisms to radiation increases their mutation rate. Muller devised clever techniques for detecting newly induced sex-linked recessive lethal mutations – mutations on the X chromosome that are lethal to the zygote carrying them. He used this technique to determine the mutagenic effects of radiation.

We must digress to explain different kinds of radiation. We are most familiar with *electromagnetic (em) radiation*, which includes ordinary light. This radiation consists of tiny packets of energy (photons), which also act like an electric and a magnetic wave moving together:

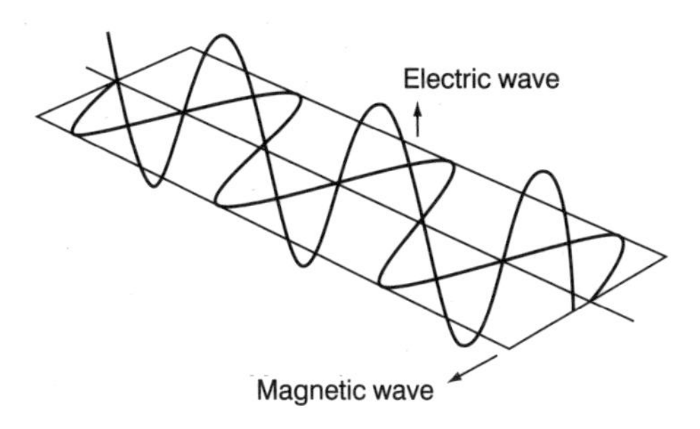

Each wave has a definite *wavelength*; the shorter this wavelength, the greater is the energy of the radiation. The electromagnetic spectrum includes visible light with wavelengths between about four hundred (violet) and eight hundred (red) nanometers:

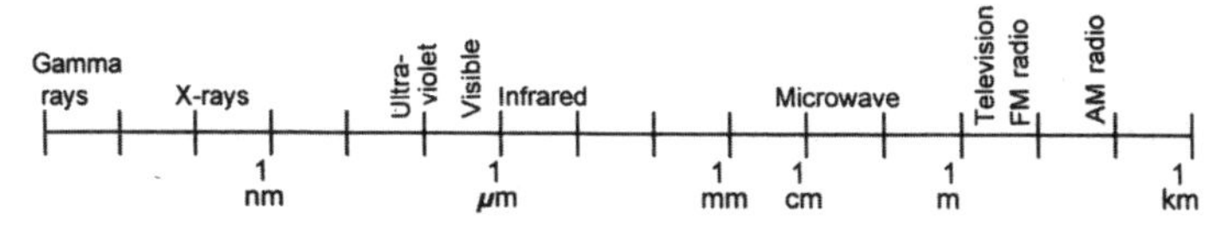

Ultraviolet light is somewhat shorter and infrared light somewhat longer. Microwave and infrared radiation can heat matter, as we do now with microwave ovens and heat lamps. Radio and television signals are carried by still longer-wave radiation, which has enough energy to move electrons through electronic circuits in our radios and TV sets so we can detect signals encoded in these waves.

Visible light has just enough energy to be absorbed by the electrons in many kinds of molecules. It is visible to us precisely because the pigments in our eyes absorb this light. Color comes from molecules that absorb light of certain wavelengths while transmitting or reflecting light of the other wavelengths. (Plants, for instance, are green because the pigment chlorophyll absorbs mostly red and blue light, whose energy is used by the plant for metabolism, but not green light.) Having absorbed extra energy, a molecule can sometimes engage in different chemical processes. Ultraviolet light (UV), which is more energetic than visible light, can sometimes cause severe chemical damage, including mutations. However, most UV light never reaches the surface of the planet, as it is absorbed by a shielding layer of ozone in the upper atmosphere. Considerable concern has been aroused by the discovery that exhausts of supersonic airplanes and the fluorocarbons in aerosol sprays break down ozone, and that a considerable part of the earth's ozone layer has already been destroyed. The possible effects of increased UV irradiation go way beyond a mere increase in human skin cancers. The use of fluorocarbons is now being curtailed by international agreements.

Electromagnetic radiation with wavelengths between about 10^{-8} and 10^{-11} meters is X-radiation. Gamma radiation is electromagnetic radiation of still shorter wavelengths that is emitted by the nuclei of certain elements. X-rays are so energetic that they knock electrons completely out of an atom or molecule, changing it into a positive *ion*. This very strong radiation is therefore called *ionizing radiation*. It can have severe effects. The free electron acts like a bullet that can leave a trail of ions as it passes through a cell, knocks electrons out of other atoms, and is eventually absorbed and held by some atom. The ions can enter into various chemical reactions. X-rays are called "hard" or "soft," depending on their energy and effects.

Particulate radiation is totally different. It consists of very energetic subatomic particles that are emitted from radioactive atoms. Beta particles are high-energy electrons. Alpha particles are a cluster of two protons and two neutrons. Few people are likely to experience much particulate radiation from a concentrated radioactive material, as in a laboratory, but we are all constantly subjected to *background radiation*, from two sources. We get a little radiation from the low concentrations of radioactive elements in the earth, and we get *cosmic radiation*, a shower of material from the rest of the universe that constantly bathes the earth and everything on it.

Cosmic radiation includes high-energy electromagnetic radiation plus particulate radiation of all kinds. Table 14.1 shows the amount of natural background radiation[1] that a typical person receives per year, compared with the additional radiation from various human-made sources, from therapeutic and diagnostic X-rays to nuclear plants and luminous watch dials. Aside from medical X-rays, the human-made radiation is quite small compared with the natural radiation.

Table 14.1. *Estimated human dosage of radiation*

Source of radiation	Average human dose (mrem per year)
Natural	
Cosmic radiation	28
Earth radiation	26
Ingested natural radiation	28
Total	82
Additional	
Medical and dental X-rays	
Patients	20
Occupational	<0.15
Radiopharmaceuticals	
Patients	2–4
Occupational	<0.15
Consumer products	4–5
Occupational	
National labs and contractors	<0.2
Industrial applications	<0.01
Military applications	<0.04
Nuclear test fallout	4–5
Commercial nuclear power	
Environmental	<1
Occupational	<0.15
Miscellaneous (air travel, TV)	<0.5
Total	30–40

what are mutations like?

A mutation is a change in DNA. Some changes occur spontaneously all the time. For instance, DNA molecules lose guanine and adenine bases (depurination) at a fairly high rate; it has been estimated that the average mammalian cell loses about ten thousand purines in twenty-four hours. Fortunately, cells have repair systems that reinsert the proper base or cut out the defective sequence and resynthesize it. But all such systems are subject to error, and such errors could well be mutations.

Various agents also damage DNA; some damages are repaired without error by cellular repair enzymes, but others can become mutations. We noted in Chapter 9 that the mutagen proflavin produces small additions and deletions of one to a few bases, and these are typically frameshift mutations, which change the reading frame that translates the genetic code. A mutation can also be a change in a single base pair. The consequences of such a change generally appear as the DNA replicates because the replacement base can form different base pairs than the original base did. The normal base pairs, A–T and G–C, are held together by specific hydrogen bonds, but the electrons and hydrogen atoms that form these bonds can shift position. A G base might shift temporarily to G*, which forms a more stable pair with T than with C. If this happened during DNA replication, it would produce a G*–T pair; then a second replication would lead to an A–T pair and thus to a mutation:

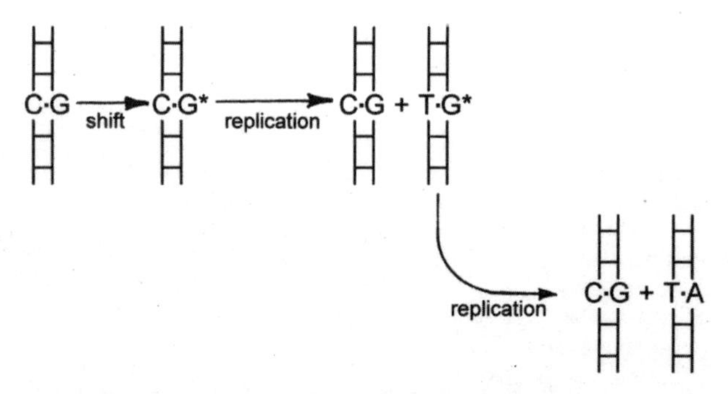

Base analogs are mutagenic molecules that are so similar to a natural base that they are incorporated into DNA. For example, 5-bromouracil (5-BU) is similar to thymine and may be incorporated into DNA as a pairing partner for adenine. However, 5-BU occasionally shifts internally so it has the pairing properties of cytosine. If this shift occurs during DNA replication, a guanine may pair with the 5-BU. One replication later, what had been an A–T base pair is changed to a G–C:

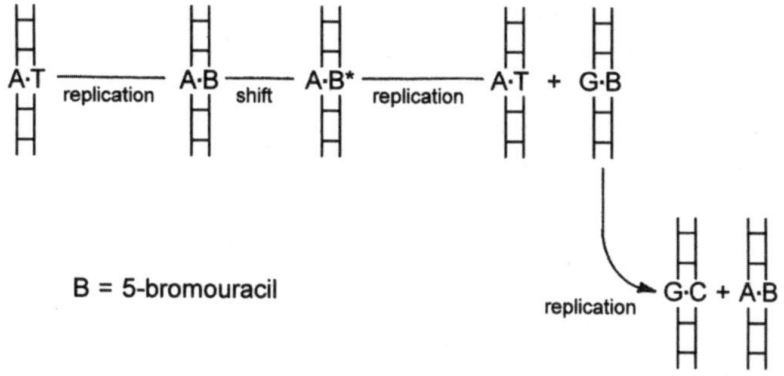

A·T —replication→ A·B —shift→ A·B* —replication→ A·T + G·B —replication→ G·C + A·B

B = 5-bromouracil

Other mutagens permanently alter bases in DNA and thus change the way they pair. For instance, nitrous acid, bisulfite, and hydroxylamine remove amino groups from bases (Figure 14.1). They convert adenine to hypoxanthine, which pairs like guanine, and convert cytosine to uracil, which pairs like thymine. Mutagens such as *nitrosoamines* add a methyl or ethyl group to a base; for instance, when guanine is converted to O-methylguanine it sometimes pairs with thymine instead of cytosine, thus producing a mutation. Nitrosoamines are commonly formed in the acid conditions of the stomach from nitrites – a serious concern because nitrites are used so heavily in cured meats. (In 1976, the U.S. Food and Drug Administration reduced the permissible level of nitrite in cured meats from two hundred parts per million to 50–125 parts per million.) These forms of damage, like

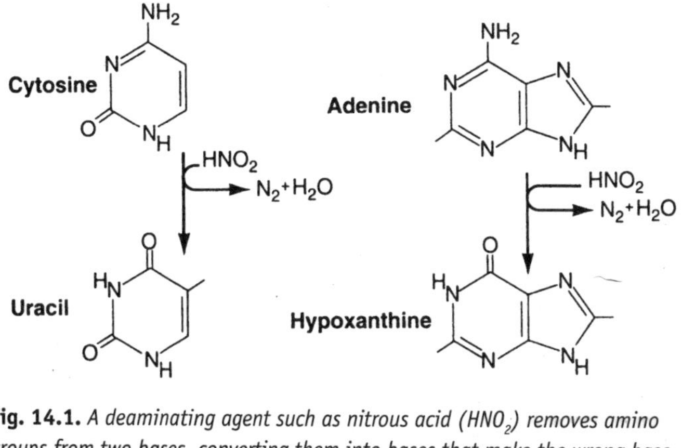

Cytosine —HNO_2 → $N_2 + H_2O$→ Uracil

Adenine —HNO_2 → $N_2 + H_2O$→ Hypoxanthine

Fig. 14.1. *A deaminating agent such as nitrous acid (HNO_2) removes amino groups from two bases, converting them into bases that make the wrong base pairs.*

those produced by 5-BU, primarily show up during replication, so dividing cells are most sensitive to these agents.

We live in a complex industrial and chemical world. So in addition to mutagens whose action is rather well known, our DNA is subject to complex and unspecified damage by various chemical agents. For instance, the combustion of many materials produces *benzopyrene,* which is activated by liver enzymes into a form that interacts with DNA; benzopyrene is made in charred meat, such as the lovely surface that a chef tries to produce in broiling a steak. Similarly, we may love peanut butter, but molds that grow on peanuts produce *aflatoxins,* another class of mutagens. This is one of the chief dangers of eating peanut butter and other peanut products. But these are only a few of the *known* chemical dangers of our world; there must be many more that we know little or nothing about.

dna repair systems

As life has evolved on this planet, cells must have encountered mutagens, both radiation and naturally occurring compounds. But it is important to keep the mutation rate within bounds, so selective pressure favored the evolution of systems to repair damage to DNA and also to correct the occasional errors that normally occur during DNA replication, for replication is not perfect.

One system of enzymes repairs damage caused by ultraviolet light, a component of sunlight. Ultraviolet light absorbed by DNA causes adjacent pyrimidine molecules in the same strand, such as thymines, to form complex structures such as a *thymine dimer*:

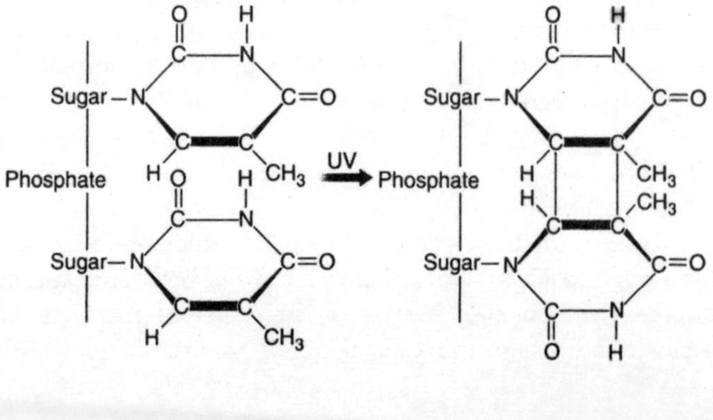

The enzymes recognize the dimer, cut it out, and replace it. Ultraviolet light is still mutagenic, however, because not all UV-induced DNA damage can be corrected. The chief health hazard is UV irradiation of the skin. Farmers, the traditional "rednecks," have elevated levels of skin cancer of the neck and other exposed skin areas, owing to constant irradiation by the UV in sunlight. Truck drivers in North America get more skin cancer on the left arm than on the right. Publicity about the long-term effects of UV light may be somewhat reversing Americans' love of the deeply tanned look; these effects include aging of the skin, so middle-aged people end up with dry, wrinkled skin that confounds the attractive look they sought through sunbathing.

Several hereditary syndromes in humans result from defects in DNA repair. *Xeroderma pigmentosum* (XP), an autosomal recessive condition that affects one in 250,000 people, is an inability to repair UV damage. People with XP are heavily freckled, very sensitive to sunlight, and develop skin cancers. Victims of another syndrome, *Fanconi's anemia,* have brownish skin, dwarfism, and skeletal abnormalities. They also produce blood cells at a reduced rate and often have leukemias, solid tumors, and many chromosomal aberrations in blood cells. The problem is an inability to repair damage to DNA from cross-linking agents. Remarkably, their cells are *less* mutable than normal cells.

genetic effects of radiation

Ionizing radiation produces every kind of mutation from point mutations to chromosome rearrangements and breaks. By placing low-level radiation sources in forests, investigators have shown that chronic exposure to radiation kills and injures plants, and thus has long-term ecological effects.

The horrifying atomic bomb explosions that destroyed Hiroshima and Nagasaki provided data on the long-term effects of radiation on people. An Atomic Bomb Casualty Commission was established by the United States and Japan to study this question. It was a huge problem to track down survivors years after the war, estimate their position relative to the center of the blast, and calculate the dose of radiation they had received. Four possible indicators of damage were considered: abnormal outcomes of pregnancy (stillbirths, major congenital defects), death in liveborn children, frequency of children with sex chromosome aneuploidy (see below), and abnormal protein variants. Survivors of the two explosions have increased chromosomal damage and cancer but, surprisingly, no significant increase in hereditary disease. It should take several generations to see an increase in diseases caused by recessive alleles or a combination of several deleterious mutations, because it takes some time to create the necessary combinations. But though induced dominant mutations should have been expressed immediately in the next generation, none appeared. So far, no hereditary effect of exposure to the bombs has been demonstrated.

Nevertheless, experiments performed with organisms of all kinds, from bacteria to mammals, show that the kind of radiation produced by nuclear weapons can cause mutations. Not only is the direct blast mutagenic, but so are particles that can be held by the winds and deposited hundreds or thousands of miles away as radioactive fallout (containing, for example, the radioactive isotope strontium-90). James Crow estimated the genetic consequences of exposing a hundred million people to an average of ten roentgens each in their gonads, an exposure that might result from a major accident at a nuclear plant. Although his calculations are necessarily crude, he estimated that such a dose would increase the incidence of dominant and X-linked disorders by considerable factors (in the range of 20–200 times), with great repercussions on the population.

Although radiation may produce cancers and other diseases in an exposed person, its most worrisome effect is on cells in the germinal tissue that produce eggs and sperm. Mutant alleles, almost all deleterious, can be passed on to future generations, adding to the genetic burden on the species. Not all of the gametes in an irradiated gonad will become mutant. Women are more susceptible to mutation in the germinal tissue, since they are born with their supply of eggs already determined and committed to release at successive ovulations; these eggs are exposed to mutagens throughout a woman's reproductive life. Men, in contrast, continually produce new sperm, and though many mutations may occur during exposure to a mutagenic agent, the sperm produced later may be normal.

Although nuclear weapons are still a great potential hazard, people are mostly concerned about the hazards from accidents or wastes from nuclear plants. The level of public concern was raised by accidents such as those at the Three Mile Island nuclear plant in Harrisburg, Pennsylvania, in March 1979, and at Chernobyl, Ukraine, in April 1986. These events have shown the fallibility of our most vaunted failsafe technology, and reports following the Harrisburg incident revealed that low levels of radiation are not as safe as had been assumed. What emerged was a chilling pattern of secrecy among government officials determined to play down the hazards of low levels of radiation. To test the combat effectiveness of soldiers close to a nuclear explosion, the U.S. Army deliberately exposed thousands to a radioactive blast and debris; a study by the Centers for Disease Control has uncovered an exceptionally high number of leukemia deaths in these soldiers. Furthermore, children born in Utah during above-ground nuclear tests in Nevada have increased rates of leukemia, directly correlated to the period of bomb testing.

A study of workers on the dockyards of the Portsmouth Naval Shipyard in New Hampshire found that exposure to radioactivity in nuclear submarines has increased their mortality and their rate of chromosome damage. Similarly, nuclear dockyard workers in Britain who had been exposed to less than the allowed limit of five rems per year for ten years had significantly increased numbers of detectable chromosome aberrations. The National Academy of Sciences of the U.S. has decided that there is no minimal threshold of biological damage by radiation.

How are we to evaluate the sensational stories that constantly arise on the subject of mutagenesis? To incidents at nuclear reactors? To the problem of storing wastes that will be radioactive for tens of thousands of years? To the entire town of Port Hope, Ontario, being radioactive from radon gas formed from nuclear wastes buried in the town by a company that once manufactured nuclear materials nearby? Although cancer creates widespread suffering and misery, the potential consequences of germinal mutation are far more serious than those of somatic mutation. We could be subjecting our species to serious, haphazard genetic change of unpredictable consequences, to elevated genetic disease, completely altering the makeup of our species – a genetic holocaust that would be virtually irreversible. Is there such a genetic time-bomb ticking away in the population? Unfortunately, the answer is that nobody knows.

chromosome aberrations

Chromosomes contain genes linked in specific sequences. A phenotype is determined by the genes that are carried and also, surprisingly, by their sequence. This seems very odd, if you think about it for a moment. Since the chromosomes are all mixed up in the nucleus anyway, we might expect that the exact location of a gene wouldn't matter, as long as it is there somewhere. So geneticists were quite surprised to find that rearrangements of chromosomes may profoundly affect the expression of the genes they carry. Just why this is so is still largely a mystery. We don't know enough yet about gene regulation, as outlined in Chapter 11, but we know that the expression of many genes is regulated both positively and negatively by DNA sequences some distance from these genes. Thus, it is possible that changing the position of a gene subjects it to different regulatory sequences. In any case, rearrangements of DNA, as well as its loss or gain, have decisive and often deleterious influences on the development of gametes and the growth of an organism.

Chromosomal aberrations result from physical breakage and reunion of ends that are not normally joined. A *deficiency*, the loss of some material, can come from a single break or from two breaks in one chromosome and loss of the intervening piece, and just rotating the intervening segment a half turn and healing the breaks produces an *inversion*:

Simultaneous breaks at different positions in two homologs can lead to a deficiency in one and a *duplication* in the other:

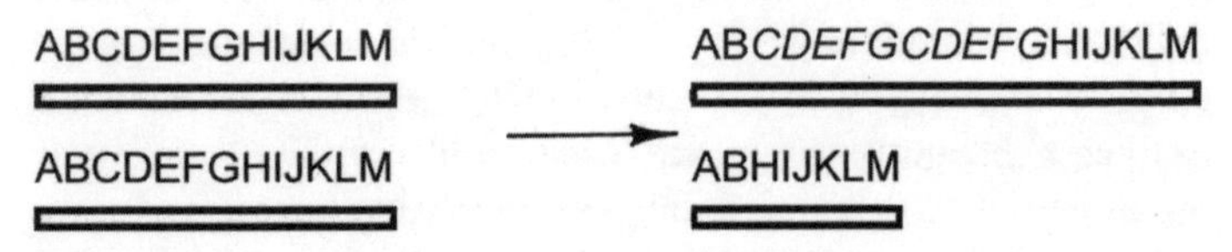

Finally, a detached piece may be transferred to a non-homolog, producing a *translocation*. (Sometimes the ends of two non-homologs exchange, making a *reciprocal translocation*.)

The number of chromosomes in a cell may also change. Instead of a normal diploid cell with $2n$ chromosomes, circumstances may add a whole chromosome set, producing a *triploid* $3n$ cell or *tetraploid* $4n$ cell. More commonly, a single chromosome is added or lost, producing an *aneuploid* organism with $2n + 1$ or $2n - 1$ chromosomes. We have already encountered aneuploidy among the sex chromosomes and have seen that it results from nondisjunction, so instead of having a pair of homologs, a diploid set may lack one chromosome (*monosomy*) or may have three homologs (*trisomy*). There is enough leeway in the functioning of the sex chromosomes so these aneuploids survive and are relatively normal, but the balance of genetic factors in the autosomes is so delicate that very few human trisomics – and no monosomics – survive gestation, and even these are grossly abnormal.

Polyploidy is common in plants, which seem to be tolerant of unusual numbers of chromosomes. There are also triploid plants, such as bananas. Many new types of plants are tetraploids that result from accidental duplication of the whole genome, and two established species commonly hybridize and produce diploid plants that carry one chromosome set from each species or tetraploids with a whole diploid set from each one. Furthermore, such new plant types may lose chromosomes and wind up with chromosome numbers that are not simple multiples of the numbers in the parental species.

It is an irony of scientific progress that the very methods of reducing infant mortality through pre- and postnatal care have also increased the chances of embryos with hereditary defects being born and surviving. Many such defects are associated with chromosome abnormalities.

Much of the sterility afflicting some ten percent of all adults owes to a chromosomal defect. Almost twenty percent of pregnant women, up to half of whom will have a detectable chromosome abnormality, will experience spontaneous abortion. At least 0.5 percent of newborn infants carry a readily perceptible chromosome aberration, and other aberrations too small to be seen easily must be common. Today it is estimated that ten percent or more of all infants are born with defects that will require major medical intervention in infancy or in later life. These numbers don't include the loss of very early embryos, which occurs before most women even know they are pregnant. So chromosome aberrations are not rare events, and they contribute to a great deal of human suffering.

looking at human chromosomes

With electron microscopes, a chromosome appears to be a thick rope twisted into many loops. Each chromosome appears to be one long, continuous fiber of DNA all folded up, combined with specific protein and RNA molecules. The study of chromosomes (called *cytogenetics*) originally concentrated on plants and insects, which have small numbers of large chromosomes; mammals tend to have numerous small chromosomes. From the 1920s to the mid 1950s, it was generally believed that humans have forty-eight chromosomes. (While D.S. was in college, he was taught that Caucasians had forty-eight chromosomes, whereas Asian males were XO with forty-seven!). But in 1956, Tijo and Levan in Sweden reported that there are only forty-six chromosomes in exceptionally good preparations of cells, and forty-six is now the accepted number for normal people. Other primates have similar numbers. Rhesus monkeys have forty-two; chimpanzees, gorillas, and orangutans have forty-eight.

We have already described (Chapter 5) how karyotypes are made by growing white blood cells with colchicine. When the chromosomes are all neatly spread out in a smear and photographed, they must be identified. All the chromosomes in a set have standard identifying numbers, starting with the longest (Figure 14.2). Next to length, the most obvious feature is the position of the *centromere*, where the chromatids are pinched together and the spindle fibers attach. Chromosomes with the centromere about in the middle, making two equal-sized arms, are *metacentric*, like numbers 1 to 3 in the human set. An *acrocentric* chromosome has arms of very unequal length (like numbers 16 to 18), while *telocentrics* have a terminal or near-terminal centromere and so have only one arm – the human set has none of these. Some chromosomes have a *satellite*, a piece connected to the bulk of the chromosome by such a thin strand that it appears to be floating near the tip.

Chromosomes can be made more distinctive by staining them in various ways. For instance, a treatment with Giemsa stain makes chromosomes develop *G-bands*, and other kinds of stain produce fluorescent bands. Such treatments help to identify the chromosomes in a set and to locate aberrations in them. In 1971, an

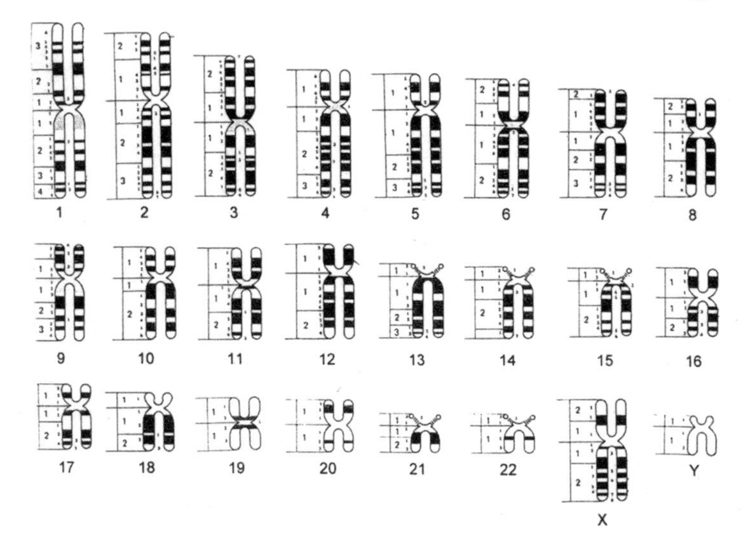

Fig. 14.2. *The human chromosomes showing the standard patterns of G-banding. The short and long arms are designated p and q, respectively, and each arm is divided into numbered segments that designate the position of each G-band. Numbering starts at the centromere. The tip of the long arm of chromosome 1 is designated 1q44 and the region next to it is 1q43.*

international conference held in Paris established a uniform system of identifying the banding patterns in each chromosome.

aneuploidy

A rich source for chromosome studies is first-trimester spontaneous abortuses, which have 50–100 times more chromosome aberrations than among newborn babies. The most common abnormality is trisomy, the presence of an extra chromosome. Trisomics for each of the twenty-three chromosomes are found among abortuses, but no monosomics except for XOs; evidently the loss of an entire chromosome impairs development so severely that the embryo dies long before it can become a fetus. Only three of the twenty-two possible autosomal trisomics survive to birth, and each has a characteristic syndrome. These are trisomy 12, trisomy 18, and trisomy 21, which causes Down's syndrome. This is the best-known trisomy, for Down's children have a distinctive appearance and are commonly seen in public these days. They are mentally deficient and for a long time were usually hidden away in institutions. More humane times and greater acceptance by society have led to their being kept at home and raised just like other children. They tend to be gentle and loving people, and they can often learn enough to hold jobs and be relatively independent.

Maternal age has a dramatic effect on nondisjunction leading to trisomy 21. The rate of Down's syndrome among eighteen-year-old mothers is only one in 2500 births, but women over forty-five at the time of conception have a rate of one in 40–50, a fifty-fold increase. Of 1700 trisomy-21 children surveyed in one study, nearly forty percent were born to mothers over forty years of age, even though only 3.5–5 percent of all normal children were born to such mothers. This age dependence is still unexplained.

Clearly, then, chromosome aberrations are a common and major source of hereditary defects in humans. That is why biologists are working to understand such aberrations, so we may at least be prepared for birth defects, even if we can't prevent them. With rapidly improving techniques of chromosome analysis, particularly higher-resolution banding techniques, carriers can be more readily recognized now. With methods to monitor the chromosome composition of fetuses already extensively used in hospitals in North America, many chromosomally aberrant fetuses are being clinically aborted. Pregnant women over the age of thirty-five are routinely screened for chromosome anomalies in many communities.

duplications and deficiencies

Just as most mutations are deleterious, large deficiencies and duplications are almost invariably lethal, and lead toaborted fetuses or grossly defective newborn babies. The best-known deficiency syndrome results from loss of part of the short arm of chromosome 5. Children heterozygous for one normal and one short-armed chromosome 5 exhibit *cri-du-chat* syndrome; they are severely defective physically and mentally and make continuous cat-like cries in infancy. Other syndromes are associated with losses of other chromosome segments, including chromosomes 4 and 18. If deficiency heterozygotes suffer from such severe defects, homozygotes for a defective chromosome must necessarily suffer from far more severe – indeed, lethal – defects. No such homozygote is known in humans, and in *Drosophila* homozygotes for almost all deficiencies die. This means that hardly any gene is dispensable, and two copies of almost all are needed for complete viability.

The events that produce deficiencies also produce duplications. The only duplications that have been found in humans are in heterochromatin, which we know little about, and in the genes for ribosomal RNA in the nucleolar organizer, which are highly duplicated anyway. It is also possible – indeed, likely – that some people carry minute duplications that are clinically insignificant and too small to be detectable cytogenetically.

The most primitive organisms have many fewer genes than do plants and animals, and we surmise that early life forms had still fewer. Duplications increase the amount of genetic material, and they are important in increasing the complexity of the genome, for in the process a cell retains all its normal genes and adds one or a few extra copies. These copies may change by gradually accumulating mutations until they may take on different functions.

Once a duplication occurs, it tends to generate additional increases in gene numbers. Again representing a series of genes by a sequence of letters:

Normal: A B C D E F G H I J K L M N O ...
Duplication: A B C *D E F G H* D E F G H I J K L M N O ...

When two duplication-bearing chromosomes occur together, they can pair unequally:

A B C D E F G H *D E F G H* I J K L M N O ...
A B C *D E F G H* D E F G H I J K L M N O ...

Then a crossover within the paired region can produce further duplication:

A B C *D E F G H* D E F G H *D E F G H* I J K L M N O ...

The result is a chromosome in which the duplicated segment is lost, leaving behind its reciprocal product, which carries a "triplication."

So once a duplication is formed, more and more genetic material can accumulate simply by asymmetrical pairing followed by a crossover. This process can go on indefinitely.

inversions

You may have personal acquaintance with couples who have repeated pregnancies and spontaneous miscarriages or abortions. Such a pattern of repeated defective fetuses may result if one parent is heterozygous for an inversion or translocation. Let's consider inversions first. Inversions can be identified by altered banding patterns; they are called *pericentric* if the inverted sequence encompasses the centromere or *paracentric* if it does not include the centromere. The genetic consequences of the two types are different.

Homologous chromosomes pair during meiosis with impressive specificity. Whatever forces hold them together are so specific that they draw homologous parts together even if the chromosomes have to twist and turn to become aligned. So during meiosis in a person who is an inversion heterozygote – that is, who carries one normal chromosome plus an inverted homolog – the two chromosomes twist into a characteristic *inversion loop* that aligns the segments between the inversion breakpoints:

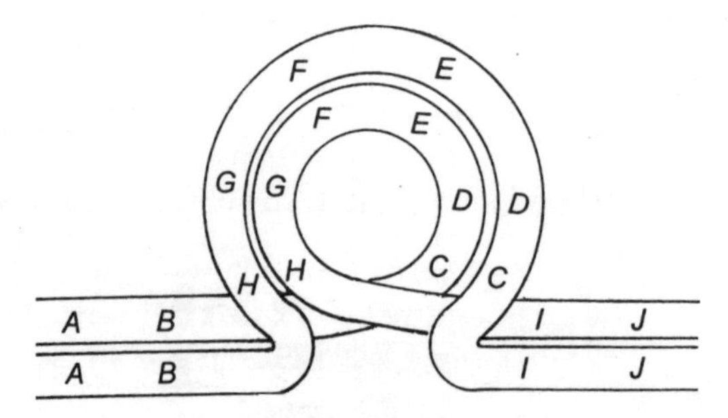

If a crossover occurs within an inversion loop, the resulting gametes may be defective. In a paracentric inversion heterozygote, crossing over hooks together two centromeres and leaves an acentric fragment with no centromere. At the end of meiosis I, the fragment is lost while the two linked centromeres remain joined by a bridge. At the end of meiosis II, the normal chromatids that were not involved in the crossover can separate; but the bridge remains, and if it breaks, the nuclei formed will have deficiencies and can never produce viable progeny. Thus the functional nuclei have only those chromatids that were not involved in the crossover. So inversions selectively eliminate crossover chromosomes. Pericentric inversions behave differently after a crossover; the chromosomes can disjoin properly, but the chromatids involved in the crossover are duplicated and deficient for the ends. Gametes carrying such chromatids have very little chance of survival. If one parent carries a large inversion, in which crossovers occur quite often, the couple will probably have an unusual number of miscarriages or give birth to malformed or mentally retarded offspring.

translocations

Translocations are a common cause of hereditary problems that we can detect through karyotyping. They are generally perpetuated by carriers, heterozygotes who carry both a translocation pair and their normal homologs but are phenotypically normal. These carriers are balanced genetically because they have two copies of all their genes, even if some genes are in different arrangements. But a proportion of their gametes are unbalanced because of events at disjunction. In such a heterozygote, the chromosomes pair in a characteristic *cross* configuration to bring homologous regions together:

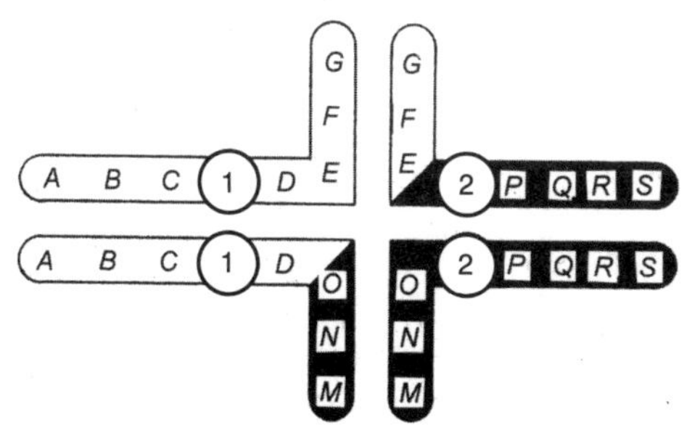

Mendel's second law still applies here. The two pairs of centromeres assort independently in two possible ways with equal frequency. If the centromeres on the same side of the cross axis migrate to the same pole, the resulting gametes carry duplications or deficiencies. If the centromeres located kitty-corner to each other go to the same pole, the gametes will be balanced with whole chromosome sets; one has completely normal chromosomes and the other bears the reciprocal parts of the translocation.

Since translocation heterozygotes generally marry people with normal chromosomes, the couple produces four types of zygotes with equal frequency: normal, normal carrier, and two kinds of duplication-deficiency products. If the duplication and deficiency are large, the fetus will probably abort spontaneously; if it develops to birth, it will probably be abnormal. For instance, Down's syndrome may result from a translocation as well as from nondisjunction. One type of gamete produced by such a heterozygote is too badly defective to produce a viable offspring. Then surviving progeny should occur in the ratio of one normal to one normal carrier to one with Down's syndrome. Thus a genetic counsellor can make specific predictions about the offspring of carriers and can advise the couple about the risks they take in having children.

chapter fifteen

evolutionary genetics

The geneticist Theodosius Dobzhansky once wrote, "Nothing in biology makes sense except in the light of evolution." All biologists who have looked at the world objectively now agree that the vast array of species on this planet has arisen through natural processes collectively called evolution. Why do they share such great confidence in this idea? Cynics and religious critics have equated this shared view with a kind of faith in the truth of evolution equivalent to the faith in God held in religion. However, the two beliefs are quite different, and this difference reflects the way science works.

Science does not deal with truth; it deals with *falsifiable hypotheses*. A scientist begins by noticing some unexplained phenomenon and trying to devise a rational explanation for it. Scientists generally reason in a pattern that the philosopher N.R. Hanson called *retroduction*: "Here is a strange state of affairs; but this state of affairs would be understandable if X were true; therefore, I postulate that X is true." Inventing an appropriate explanation requires a leap of the imagination, and for this reason science is as creative an activity as the arts or any other human activity. However, science imposes several limitations on what X might be like. X must be in the realm of ordinary physical reality; a scientist cannot postulate something supernatural or in principle unobservable, such as a god or demon or some kind of magic. But most important, the postulate about X is a hypothesis that must be subject to an *empirical* test – it must have consequences that are testable through observation and experiment. Then we challenge the hypothesis by testing some of these predictions.

Non-scientists tend to imagine that in doing experiments or making critical observations, scientists are trying to prove that their hypotheses are correct. In fact, they are trying to prove that their hypotheses are wrong. Of course, scientists hope they are correct, but a quirk of logic dictates that they can't try to prove this. A hypothesis, *H*, and its predictions, *P*, form a hypothetical statement of the form, "If *H*, then *P*." So we test to see whether we do observe *P*. Suppose we do. Can we reason, "If *H*, then *P*; *P* is true, therefore *H* is true"? No, we can't. This pattern of reasoning is a logical fallacy called "affirming the consequent." (Try reasoning, "If the sun is made of burning cow dung, it will be hot; the sun is hot, therefore it is made of burning cow dung.") But suppose *P* is not true – that we observe not-*P*. Then we *can* reason, "If *H*, then *P*; not-*P*, therefore not-*H*." As the philosopher Karl Popper emphasized, the hypothesis must be *falsifiable* – in principle there must be some empirical test that will challenge it. If the results of a test do not support the hypothesis, it is rejected or at least modified. If the results do support the hypothesis, it survives, and we gain greater confidence in it. This does not make the hypothesis true, merely an acceptable explanation for the observations.

As we develop more hypotheses that survive many empirical tests, we can piece together larger explanatory structures called *theories*. However, even theories are not truths; they simply become more and more robust and more all-embracing explanations of natural phenomena. Evolution is generally called a theory of this kind, one that has survived many challenges and has thus become very strong. But we must be careful to distinguish two aspects of the idea of evolution: the question of whether evolution *does* occur (have existing species acquired their present form in this way?) and the question of *how* evolution occurs. That evolution does occur is so well founded that denying it would make nonsense of virtually all of biology, as Dobzhansky observed; mountains of evidence and experimental tests, over the past 150 years or so, have supported (and hence failed to falsify) the general idea that species evolve and that evolution explains the diversity of species.

The second aspect of evolution – accounting for its mechanism – is still a topic of scientific controversy, like many things in science. However, there is universal agreement that evolution depends upon *natural selection*, as explained below. Credit for the idea of evolution through natural selection is given to Charles Darwin and Alfred Russel Wallace, although many scientists had entertained evolutionary theories before them. Both men traveled the world extensively, carefully observing and recording the wide range of exotic plants, animals, and fossils they saw. In this diversity, they both saw clear patterns of similarity. Darwin made the point using the example of vertebrate forelimbs with their common structure of one long bone in the upper part, two in the lower, several bones in the wrist area, and five digits made of a series of short bones:

> What can be more curious than the hand of a man, formed for grasping, that of a mole for digging, the leg of a horse, the paddle of a porpoise, and the wing of a bat, should all be constructed on the same pattern and should include similar bones, in the same relative positions?

This common structure in the face of varying functions is called *homology*. Anatomists before Darwin had seen homology and generally explained it with reference to traditional religious ideas, though sometimes toying with evolutionary notions. Darwin and Wallace explained homology by postulating descent with modification from a common ancestor. They proposed that ancestral stocks of organisms must have split up into populations that were subjected to different environmental influences. In each population, any organisms with slight inheritable changes that made them better suited to that environment would be *naturally selected* over the unchanged ones. Through repetition over many generations, random changes in genomes followed by selection would lead to progressive divergence of types, and ultimately to the modern forms such as human, bat, and whale in the example above.

evidence for evolution

The evidence that organisms have really evolved from common ancestors, with gradual modification in each line of descent, comes from several sources. Homology at all levels of biological organization is probably the strongest evidence. Studies of all groups of organisms have piled up extensive examples of homology in visible morphological structures such as limbs and skulls. Studies of homology extend from comparisons of currently living species to an extensive range of fossils. We can identify these fossils as the ancestors of modern species, filling in gaps and confirming relationships that can only be inferred from the living forms. Putting all this information together, we can summarize the evolution of each group as a *phylogenetic tree*, a branching pattern showing how each species is related to its ancestors. The fossils can also be dated by the rock strata they lie in, which provides a realistic time scale for evolution.

More recently, our ability to sequence proteins and large amounts of DNA from many species has revealed homology at the molecular level. First, there is an enormous overlap of gene content between organisms: a certain gene in one organism is very likely to have a counterpart in some other organism. (This has given a great boost to the study of model organisms, because what is discovered in the model can probably be applied to other species.) Second, homologous genes show great similarities in nucleotide sequence (or amino acid sequence of their protein products). The sequences of homologous proteins – for instance, all the insulins or the hemoglobins of vertebrate animals – are very similar, with minor differences. By analyzing the pattern of differences, we can infer in what order these differences probably arose over the long periods that these organisms have been diverging from one another, although a computer program may be needed to cut through the complexities. When this information is made into hypothetical trees of evolution, showing the most likely series of changes, the result generally confirms the phylogenetic trees that have been drawn on the basis of obvious anatomical similarities; sometimes information about molecular structure clears up an old mystery or reveals relationships that anatomy could not show.

Third, in closely related species the genes are often in the same positions relative to other genes, a situation known as *synteny*. The degree of synteny is greater within a group of related organisms, such as mammals, than between more distantly related groups.

Another type of evidence for evolution is that we can observe progressive genetic changes in populations over historical time. Many examples have been documented of such changes occurring in response to environmental changes caused by humans, including the rise of antibiotic-resistant bacteria; rats resistant to the poison warfarin; rabbits resistant to the disease myxomatosis, which had been used to exterminate them in Australia; and plants resistant to heavy metals contaminating soil. One classic study showed how populations of a light-colored British moth (*Biston betularia*) became very dark in industrial regions where the trees turned dark from pollution (which killed the light-colored lichens on the bark), so they were well camouflaged. As the pollution was eliminated, the lichens grew again, the trees turned light again, and so did the moths.

Cases of more "natural" evolution have also been observed. A good example is the adaptive radiation of *Anolis* lizards introduced into certain islands of the Caribbean. And Peter and Rosemary Grant have observed that populations of ground finches (Geospizinae, now called "Darwin's finches") of the Galapagos Islands undergo heritable adaptive changes, as in the shapes of their bills, in response to severe environmental changes such as periods of drought and extreme storms. These small changes show how even very small genetic differences can be adaptive.

evolution as a process

From a broad perspective, evolution entails three processes: macroevolution, speciation, and microevolution. *Macroevolution* means the large-scale changes observable in the fossil record, which shows that vastly different types of organisms have existed in the past and have replaced one another right up to the present. *Speciation* means the formation of two or more distinct species in place of one ancestral species. *Microevolution* means much smaller changes within a single species. This naturally raises the question of what a species is. There is no simple answer. It clearly means a "kind" of organism, but a glance at any book about trees, flowers, birds, insects, or any other organisms will show that many forms designated distinct species are very similar to one another. So how are the lines to be drawn? The simple (and highly debatable) answer is that for sexual organisms the line is drawn at reproduction: all the members of a single species are potentially capable of breeding with one another, but they cannot reproduce with members of other species. Sometimes the members of a species can *look* very different but can still interbreed. Domestic dogs are an impressive example: the largest types, such as Newfoundlands, can breed with the smallest, such as chihuahuas, to produce fertile offspring.

Speciation is a critical process in evolution. It is the step by which the organisms that were formerly all one species – all able to interbreed with one another – divide into two or more groups that are *reproductively isolated.* From that point on, each species is able to undergo an independent evolution. Thus, it is speciation repeated again and again over the eons that has produced the highly branched tree of life, extending today to untold millions of living species. Speciation occurs in various ways. A considerable amount of evolution in plants, for instance, occurs through accidents and hybridizations that enlarge a set of chromosomes. For instance, the evolution of wheat started in ancient times, over ten thousand years ago, with the combination of two diploid ($2n$) species to produce a tetraploid ($4n$) that is still known as Emmer wheat. Then around eight thousand years ago, hybridization with yet another species produced the hexaploid ($6n$) form that became modern wheat. In another type of plant evolution called *introgressive hybridization*, two plants combine their genomes in a cross, followed by the formation of a new, viable plant that has a few chromosomes from one ancestor combined with the bulk of the genome of the other.

However, it is clear that much speciation, especially in animals, requires *geographic isolation.* A large species may spread out geographically over a wide area. Within that range, it acquires a degree of geographic variation, which means that forms in different areas acquire distinctive features. They often become distinct enough to be named separate *subspecies*, or *races.* (This is obvious in many species of birds, so birdwatchers have to learn these differences as they visit different areas.) With this much variation already in the species, some populations may become isolated by barriers – glaciers, rivers, plains that interrupt a forest, and so on. During the time of isolation, the separate populations pick up additional genetic differences, including *reproductive isolating mechanisms* that prevent them from interbreeding when they eventually come back into contact. Some possible mechanisms may be incompatibilities between chromosome sets, barriers to fertilization, different times of breeding, and the inviability of hybrid offspring.

Darwin first observed evidence for this kind of speciation in the ground finches of the Galapagos Islands off the coast of Ecuador.

Island groups can be hotbeds of speciation. A variant population from one island migrates to another island, where they undergo a separate evolution for a while. They may expand their range later, but during their isolation they have become different enough to remain a distinct species.

Looking at macroevolution overall, we see at least three kinds of events. Speciation occurs over and over. We also see single species gradually changing over a long period, a process called *phyletic evolution*. But a dominant feature of evolution is *extinction*; evolution is not so much a story of perpetual survival but, rather, of temporary survival ending in failure. Species tend to survive for various periods, ranging from a few hundred thousand to a few million years, and then to become extinct when they are unable to adapt to changing environmental conditions (which include other organisms that are also evolving).

population genetics

Talking about wild-type alleles and mutant alleles, as we have done to introduce basic genetic concepts, is simplistic and misleading. Studies of real populations show that the individuals comprising them do not all share a single genotype that we could call the wild-type; in fact, natural populations show enormous genetic diversity. Dobzhansky and his associates studied wild *Drosophila* populations in the American southwest and showed that they carry several distinct *inversion* variants of each chromosome. Remember that in an inversion a segment of a chromosome is turned around. In their salivary glands, fruit flies have giant chromosomes, characterized by distinctive black and white bands that can be seen easily with a good microscope. It is easy to compare the banding patterns of chromosomes from different individuals and see where one has an inversion relative to another. A key concept of population genetics is *allele frequency*, which simply means the fraction of a certain gene or chromosome type in a population. For instance (using names from Dobzhansky's work), suppose thirty-seven percent of a certain population of fruit flies have number 2 chromosomes with the Standard gene sequence, sixteen percent have the Arrowhead inversion, and forty-seven percent have the Chiricahua inversion;

then the frequencies of these forms are 0.37, 0.16, and 0.47, respectively. Dobzhansky and his colleagues mapped the frequencies of different inversions across the whole region and showed that the frequency of each inversion type changes regularly from California eastward and down into Mexico, presumably because certain gene sequences give their bearers greater selective value in each geographic region. Other studies of wild populations show similar results. Many genes and chromosomes have allelic variants, and they remain in the population at substantial frequencies, perhaps even changing regularly with the changing seasons. This enormous variation is the source for evolution.

This variation arises because mutations continually occur at low rates throughout a population, changing genotypes at random. Some of these random changes turn out to be adaptive in the organism's environment, so individuals that have them produce more offspring than do unmutated individuals. Because of their reproductive superiority over the generations, the proportion of mutant individuals in the population increases progressively over time. Natural selection simply means this kind of differential reproduction. Every genotype has a certain relative *fitness*, measured by its reproductive rate; saying that a certain genotype has a high fitness or is naturally selected just means that it is more successful than other genotypes in leaving copies of itself in succeeding generations.

To produce new species or higher taxonomic groups such as genera, changes are required at many gene loci. As a simple example, we might imagine that an adaptive change in some species might be a genetic change from *AA BB mm QQ stst* to *aa bb MM qq StSt*. For this to occur, mutations would have to occur from *A* to *a*, *B* to *b*, *m* to *M*, *Q* to *q*, and *st* to *St*. The mutations would occur at different times, repeatedly, and in different individuals, with the final genotype assembled by recombination. We can imagine mutations sculpting the vertebrate limb to lengthen, thicken, or shorten certain bones, leading to the patterns we see today. Some investigators have simulated the selection of certain genotypes in the laboratory.

Population genetics attempts to describe this process quantitatively. It begins with a model for one gene. Suppose a population has alleles *A* and *a* of a single gene and that the frequency of *A*, represented by *p*, is 0.6 and the frequency of *a*, represented by *q*, is 0.4. (Note that in this simple model, $p + q = 1$, because all the alleles in the population must be either *A* or *a*.) We can determine these frequencies by counting individuals, both homozygotes and heterozygotes. Every homozygote carries two copies of the same allele, and every heterozygote carries one copy of each.

What will the genotype frequencies be in this population? The twin processes of mutation and selection act slowly across many generations, and to begin we assume they are not acting at all. We also assume that the population is large enough for the laws of probability to apply to it, and that individuals mate at random. This means that neither males nor females select their mates in any way; an *AA* individual does not prefer to mate with others of the same genotype, for instance. Now remember that the gametes will carry either *A* or *a*, but not both, so the gametes of the two types will also occur in the frequencies p and q, respectively. The gametes (alleles) in a population are analogous to a bag full of marbles, some red (*A*) and some blue (*a*). To make an individual, we reach into the bag blindly with both hands and withdraw two marbles. The probability of them both being red is $p \times p = p^2$, and the probability of them both being blue is $q \times q = q^2$. Sometimes the left hand will pick a red and the right a blue (frequency $p \times q = pq$) and sometimes the other way around ($q \times p = qp = pq$). So the frequencies of the genotypes should be: p^2 for *AA*, $2pq$ for *Aa*, and q^2 for *aa*.

This approximation, called the Hardy–Weinberg formula, is the foundation of population genetics. It has been developed in sophisticated mathematical detail to analyze the process of evolution by assigning values to rates of mutation and the selective advantage an allele may enjoy. Another application is in understanding human population structure regarding single-gene diseases. For example, phenylketonuria is a disease of autosomal recessives, which occur in the population at a rate q^2. If one person in ten thousand in a certain population has phenylketonuria, $q^2 = 1/10{,}000$. So q must be the square root of 1/10,000 or 1/100. Since $p + q = 1$, p must be 99/100. Then according to the Hardy–Weinberg formula the frequency of heterozygous carriers is $2pq = 2 \times 99/100 \times 1/100 =$ (approximately) 1/50. This interesting calculation shows that heterozygotes are vastly more common (here, one person in fifty) than those who have the disease. The heterozygote frequency is important for genetic counseling. It also has important implications for the ability to eliminate the recessive allele from the population by selection, as we shall see later.

human evolution

The most controversial of Darwin's ideas about evolution was his assertion that humans had evolved from apes. The resistance to the idea came from the Christian church because the idea of evolution was at odds with the generally accepted biblical story that all species on earth had been divinely created during a brief period. At the time there was little evidence to support Darwin's contention about human evolution, beyond a general resemblance of the body plans of humans and apes. Anthropologists began to search for "missing links." fossil forms that could be intermediates between apes and humans. In the subsequent 150 years, several distinct types of intermediates have been found, so anthropologists can draw a reasonable picture of the evolution of humans and human ancestors, collectively called *hominids*.

The oldest presumptive intermediates have been found in Africa, making it virtually certain that hominids originated on that continent. The earliest hominids that are fairly well known have been dated geologically to around 4–3.75 million years ago (My). Named *Australopithecus afarensis*, these hominids were bipedal, quite short (around 1.2 meters), and had small brains. They were succeeded by two kinds of *Australopithecus*, named *Australopithecus africanus* and *Australopithecus robustus*, that were larger and had bigger brains. These later australopithecines apparently coexisted around 2.5–1.5 My with the first forms assigned to the genus *Homo*: *Homo habilis*, named from the Latin word for "skillful" and nicknamed "handy man," because they were able to make simple tools. *H. habilis* people were somewhat taller (1.5 meters) and had larger brains, about half the volume of modern humans. They were replaced by the taller, larger-brained *Homo erectus*, which lived around 1.0–0.25 My. From 250,000 years ago to the present, the fossils are of modern humans, generally called *Homo sapiens*. However, most of the earlier forms, before about thirty-five thousand years ago, are Neanderthal humans, which have been called either a subspecies of *H. sapiens* or a distinct species. Recently the British geneticist Brian Sykes has used

DNA evidence to argue that the Neanderthals have left no trace in the modern human genome and were a distinct species, *Homo neanderthalensis.*

the migration and diversification of *homo sapiens*

Studies of the DNA sequences of humans from all over the globe have been used to produce phylogenetic trees of humankind. These trees are rooted in Africa, from the fossil evidence. Most biologists believe such trees support the "Out of Africa" theory, that *H. sapiens* evolved in Africa and then migrated by several routes to all parts of the world. Divergence among groups during and after this migration produced a variety of so-called human races, if a race be defined as "a visibly distinct group of the human species, originally found in one geographical area."

Since the history of contacts between diverse people shows that all humans are capable of interbreeding, we all belong to a single human species, *H. sapiens*. This might seem hard to believe if we compare very different-looking peoples such as the short, dark-skinned pygmies from central Africa and tall, light-skinned northern Europeans. Anthropologists have used these obvious differences to distinguish races, in an effort to catalog human variation. However, different anthropologists had different ways of defining races, so there was no universally agreed classification. The numbers of races proposed ranges from as few as four ("black," "red," "yellow" and "white" skinned) to as many as fifty. One classification proposed seven main races: Caucasians, Black Africans, Mongoloids, South Asian Aborigines, Amerindians, Oceanians, and Australian Aborigines. All this indicates a great deal of genetic variation within races – variation that allows us to distinguish individuals.

The visible surface characteristics used to define races – size, skin color, eye shape – are genetically determined. But all human characteristics show variation, and the members of each group can be quite diverse in appearance. For example, "Caucasians" include light-skinned northern Europeans, darker-skinned Mediterraneans, and dark-skinned nothern Indians. DNA sequences can provide a better indication of variation, as measured by the number of different forms of individual genes (or individual segments of DNA) in a population. DNA sequencing shows that there is more variation *within* races than *between* races. But races *seem* very different, so how can this be true? It is true because measures of DNA variation include *all* genetically determined variation, not just the visible features originally used to characterize races. An analogy is having a set of a hundred black beads, all of different sizes, shapes, and textures, and another set of a hundred white beads also of many sizes, shapes, and textures. There is much variation within both sets, and the only variation between the sets is in a single conspicuous feature: color.

The DNA evidence confirms that the concept of race is quite arbitrary. Although race has figured prominently in human history, biologically it is "only skin deep." At the genetic level, there are no significant boundaries between races other than in the gene coding for superficial properties like skin color. So at the genetic level the so-called races blend into one another, and a scientist presented with a particular DNA sample would probably be unable to tell which race it came from.

The routes of migration out of Africa have been traced by examining the patterns of loss of genetic diversity. Assume that the population in some region – call it region 1 – has considerable genetic diversity, including DNA forms a, b, c, d, e, and f. In a nearby region 2, the population has only forms a, b, c, and d, and in nearby region 3 the population has a and b only. This pattern has been interpreted by means of the *founder effect*: that a small group that become the founders of a new population happen to carry a set of genes that is atypical of the larger population they come from. In this case, we infer that a small group of people from region 1 migrated to region 2 and established a population there. However, these founders did not carry forms e and f. Later a small group lacking c and d left region 2 and founded a new population in region 3. In this way, migration routes have been mapped throughout the continents. For example, one major route swooped into southern Asia, and then a branch went north, spiralling back into central Asia.

Brian Sykes and his colleagues have been tracing human migration patterns through mitochondrial DNA. Mitochondria contain a small, distinctive DNA molecule that encodes several proteins of this organelle; it is easily amplified with PCR and sequenced, so a person's mitochondrial DNA can be characterized from a small blood sample and even from some fossil remains. But mitochondria are inherited strictly through a maternal line; the sperm carries only a male nucleus into the egg, but no mitochondria. Using this method, Sykes was able to settle the longstanding issue of whether the Polynesian islanders came from Asia or from South America; their source was unequivocally Asian. By collecting samples throughout Europe, Sykes has identified seven ancestral centers from which mitochondrial DNA has spread in European populations – meaning, in effect, seven hypothetical women who lived between about forty-five thousand and fifteen thousand years ago.

Note that modern Africans have been changing genetically in the same way that all their descendants in other parts of the world have been changing. Modern Africans are not the same as the original African stock from which the migrations started. Modern Africans are in no sense primitive.

The visible distinguishing features of races arose by small-scale DNA changes that accumulated differently in different locations on earth. It seems likely that many of these changes were adaptive, that they better equipped people for life in the environments they migrated to. Other genetic changes may have accumulated by chance through *genetic drift*. The Hardy–Weinberg law operates in a large population because many matings occur at random, but in a small population the genetic coins are being flipped only a few times, so gene frequencies may drift rapidly just by chance. Still other changes may have accumulated as a result of the founder effect; for example, many of the founders might have red hair, while only one in ten people in the original population have red hair, so the new population can be quite different from the original population.

Some of attributes that vary geographically probably have real adaptive value:

skin color

Average skin darkness is strikingly correlated with latitude; in many regions we see a gradient from dark skins near the equator to light skin closer to the poles. Dark skin may provide protection against ultraviolet rays in sunlight, which cause skin cancer and are more intense near the equator. On the other hand, vitamin D is synthesized in the skin by the action of sunlight; the lighter skins of more northern people permit more sunlight to enter, to promote synthesis of vitamin D, while darker, equatorial skins help prevent excess synthesis of vitamin D, which can be harmful.

body shape

A slender body shape is more effective at losing heat, and examples abound in equatorial peoples. A rounder shape, typified by the Arctic Inuit people, conserves heat better, a potentially useful attribute in colder regions.

resistance to malaria

Heterozygotes for sickle cell anemia are somewhat resistant to malaria, so the mutant allele of the hemoglobin gene that causes this condition is correlated geographically with the presence of the malaria parasite (mostly in the tropics). Although homozygotes die from the condition, many benefit from carrying this allele.

survival at high altitude

People who have lived at high altitudes, as in the Andes mountain range, have various physiological adaptations that make them better able to perform with reduced concentrations of oxygen.

eugenics

As we noted in Chapter 1, the idea of improving the human species is at least as old as some classical Greek societies. But the idea took hold strongly in some quarters in the late nineteenth and early twentieth century in the wake of a popularization of Darwinian ideas about natural selection, and the discovery of single gene inheritance. The idea of improving the human species by social action was first formulated by Darwin's cousin Francis Galton, who termed the practice *eugenics*, literally meaning "good birth." Galton reasoned that the same genetic principles we use to improve animals and plants ought to be applied in the selective breeding of humans. The concept was espoused by many intellectuals of the time, including the writer George Bernard Shaw. In both Europe and North America, social programs of eugenics were implemented. Eugenicists attempted two broad applications, one focused on medical conditions and the other on the alleged superiority and inferiority of certain races or social groups.

Many human diseases and other phenotypes had been shown to be inherited as single alleles, according to standard Mendelian principles. However, eugenicists alleged that other, more dubious conditions were also inherited thus, such as the ill-defined states of "feeble-mindedness," "nomadism," and "criminality." Eugenicists set out to eradicate these conditions from the population. In the U.S., Canada, and several European countries, eugenics boards were set up to assess individual cases, and as a result tens of thousands of people were sterilized, many for no sound reason. In some countries, such as Canada and Sweden, these practices continued until the 1970s. Indeed in 1995 the People's Republic of China passed the "Maternal and Infant Health Care Law" to try to reduce the incidence of genetic disease, a law that many considered to be eugenic in nature.

There are no data to suggest that any of these programs had an impact on the incidence of people with disorders. For recessive disorders, this is not surprising; selection against recessive disorders is inherently inefficient because most of the recessive disease alleles reside in heterozygotes. From population-genetic principles, it is easy to calculate that if all those with a recessive condition are kept from breeding (the most severe form of selection), to halve the frequency of a typical disease allele from 1/100 to 1/200 would take a hundred generations or about two thousand years. Selection against a dominant condition would be more effective, but even here problems of variable gene expression reduce the efficiency of the process.

The history of the world shows many examples of belief in the superiority or inferiority of certain races or social strata. These erroneous notions have been and still are the cause of much human suffering. During the early part of the twentieth century, when North America was experiencing its peak of immigration, eugenic logic was applied to restrict the entry of certain races and social classes on the grounds of inferiority. In the U.S., the Cold Spring Harbor Laboratory, now a prestigious biological research institution, was originally founded as a Eugenics Record Office, focusing in large part on immigrants. Additionally, society was riddled with spurious notions about the genetic basis of social standing, and it is not surprising that eugenics found favor among the privileged, who believed that a proper eugenic program would produce a society of people just like them. One account lauded the statistics that it takes on average forty-eight thouand parents who are unskilled workers to produce a single child who is written up in *Who's Who?*, whereas only forty-six of the professional class are needed. This was considered support for a genetic meritocracy.

In Germany, eugenics began slowly but spiraled out of control. A 1933 law aimed to eliminate progeny with hereditary defects by stipulating compulsory sterilization of people with congenital mental defects, schizophrenia, manic depression, hereditary epilepsy, and severe alcoholism. This program was set up with the compliance of several senior German geneticists. In 1937, children of color were added to the sterilization list, and in 1939 an official program began for the euthanasia of psychiatric patients. In 1942, the mass murder of Jews and Gypsies began in the concentration camps. These atrocities all served a notion of creating some type of pure "master race" by cleansing the population of "inferior" alleles. The Germans also tried selective breeding of those with the favored "Aryan" characteristics.

With modern genetic knowledge, some communities have now undertaken a voluntary program to reduce the frequency of a specific genetic disease in that community. For example, Tay–Sachs disease is an autosomal recessive lethal disease of the nervous system; the causative allele has a high frequency among Ashkenazi Jews, originally from eastern Europe. The disease was common among the newborn in many immigrant Jewish populations in North America. However, prospective parents can now be screened with DNA diagnostic techniques to identify heterozygotes for the Tay–Sachs allele. Couples who are both heterozygotes often choose not to marry. In this way the disease has been almost eradicated in some communities, as in New York City. However, let us emphasize the important distinction between voluntary participation in such a program and enforced participation in a program imposed by the state.

It is worth noting that DNA diagnostics are available for detecting many recessive alleles in the heterozygous condition, and no doubt many more will be developed soon. So it may become possible for more extensive voluntary screens for other diseases, such as cystic fibrosis, to be carried out on the population.

The ethical and the practical sides of eugenics need to be distinguished. Most historical applications of eugenics were ethically and scientifically flawed. They were scientifically flawed partly because of incorrect understanding of the hereditary basis of the conditions selected against and partly because of the inefficiency of selection against recessive conditions. However, the rapid advances being made in genetic research may well remove both these practical objections. As overpopulation, global pollution, unstoppable infectious diseases, and resource shortages loom, it is conceivable that one day humanity might be forced to consider taking a godlike control of its own evolution.

glossary

Some terms are distinguished as adjectives (adj.), nouns (n.), or verbs (v.).

active site The spot or cavity on an enzyme molecule where a chemical reaction takes place.

adenine One of the purine bases of DNA or RNA.

agar A polysaccharide derived from seaweeds, used to make semi-solid nutrient media.

alcaptonuria A metabolic disorder characterized by mental retardation, indicated by secretion of homogentisic acid (alcapton), which becomes dark on exposure to air.

alcohol An organic compound bearing the hydroxyl group, OH.

allele One alternative form of a gene.

allele frequency In population genetics, the fraction that a given allele represents out of all the alleles of a single gene (or chromosome type).

allosteric protein A protein that has two distinct binding sites, each specific for a different ligand, and which changes conformation and activity as it binds to these ligands.

***amber* mutant** A mutant whose defect is a nonsense codon (UAG), which terminates protein synthesis prematurely.

amine An organic compound bearing the amino group, NH_2.

amino acid A monomer of a protein, an organic compound that has both an amino group (NH_2) and an acid (carboxyl) group (COOH).

amino group the chemical combination NH_2.
amino terminal The end of a peptide chain with a free amino group.
aminoacyl tRNA A transfer RNA molecule combined with the amino acid it carries for use in protein synthesis.
anaphase The phase of mitosis or meiosis in which chromosomes or chromatids separate and move toward opposite poles.
aneuploid Having a chromosome set with extra chromosomes or deficient in chromosomes, relative to the normal (euploid) set.
antibody A molecule made by certain cells (generally in a bird or mammal) in response to an antigen, one that is generally able to combine with the antigen and neutralize it.
anticodon The sequence on a transfer RNA molecule that is complementary to a codon sequence in a messenger RNA molecule.
antigen Any foreign material introduced into a body (typically of a bird or mammal) which induces the formation of antibodies against it.
autoradiography A technique of exposing radioactive materials to a photographic gel so the location of the radioactive label is shown by dark spots when the gel is developed.
autosome An ordinary chromosome, as distinct from a sex (X or Y) chromosome.
auxotroph A mutant organism that is unable to synthesize one or more of its essential components.
bacteriophage A bacterial virus.
base (1) A substance that accepts hydrogen ions. For instance, sodium hydroxide (lye) is a base because the molecule, NaOH, comes apart in water to form an Na^+ ion and an OH^- ion; the latter combines with a hydrogen ion, H^+, to make another water molecule. (2) Specifically, one of the large, nitrogenous components of a nucleic acid.
bioinformatics The science of analyzing the structure and function of a genome, especially by means of computer models and programs.
biosynthesis The part of metabolism responsible for synthesizing the molecules of which organisms are made.
blastula The stage of a developing animal embryo consisting of a hollow ball of cells.

breed true To consistently produce offspring of a given type.

capsid The protective protein coat of a virion (virus particle).

carboxyl group The chemical combination COOH, which is an acid group because the hydrogen atom tends to detach as a hydrogen ion, H^+.

carboxyl terminal The end of a polypeptide chain that has a free carboxyl group.

catalyst A substance that speeds up a chemical reaction but is not permanently consumed in the reaction.

cDNA See *complementary DNA*.

cell A fundamental biological unit, consisting of a highly organized complex of biological molecules and ions in water, all surrounded by a plasma membrane.

cell cycle The cycle of events, including DNA replication, that results in division of a cell into two cells.

cellulose A polymer of glucose, which forms an important part of plant tissues, especially wood.

centriole A structure in animal (and some other) cells that resides at one pole of a dividing cell and helps to direct the movement of chromosomes toward that pole.

centromere The point on a chromosome where the chromatids are attached to each other.

chase To add non-radioactive material to cells after giving them a radioisotope, in order to flush the radioactive material on to its destination and remove unincorporated radioisotope.

chemical bond An interaction between two atoms that keeps them attached to each other (see *covalent bond*).

chemical reaction A process in which atoms or molecules interact with one another and rearrange themselves to form different molecules.

chiasmata/chiasma During meiosis, an X-shaped interaction between homologous chromatids during which they may break and exchange parts with each other.

chloroplast An organelle of photosynthetic eucaryotic organisms, such as plants, in which photosynthesis occurs.

chromatid In a chromosome that has been replicated, one of the two identical halves that constitute the whole chromosome.

chromomere A small bulge on a chromosome.

chromosomal aberration A change in chromosome structure or number.

chromosome A structure consisting of DNA complexed with histones and other proteins and with RNA molecules; one of the principal structures that constitute the genome of a cell.

cistron A functional unit essentially equivalent to a gene, but defined on the basis of a complementation test.

clone (1, n.) The set of all organisms derived from a single original organism through asexual reproduction, such as cell division. (2, n.) An organism created by inserting the nucleus of an organism to be copied into the enucleated egg of another organism. (3, n.) a section of DNA that has been inserted into a vector and replicated to form many copies (*DNA clone*). (4, v.) Experimentally create a clone, in the sense of (2) or (3).

coding strand The strand of DNA whose base sequence is identical to that of a messenger RNA, except with thymine instead of uracil.

codominant Two alleles are codominant if they are expressed equally in a heterozygote bearing both of them.

codon A unit (of three nucleotides) that encodes a particular amino acid.

colchicine A substance that interferes with mitosis and allows dividing cells to stop in metaphase, in order to make a karyotype.

cold sensitive Adjective referring to a mutant that functions normally at mid-range temperatures but is abnormal at low temperatures.

colinear The relationship between a nucleic acid and the protein it encodes whereby the nucleic acid sequence directly specifies the amino acid sequence.

colony A mass of cells, such as bacteria, that have grown as a clone from one original cell, generally on the surface of nutrient agar.

complement The nucleotide bases that form stable pairs with one another – A with T or U, G with C – are said to be complements of one another. Two nucleic acid strands with complementary base sequences are also complements.

complementary Molecules whose shapes fit together, allowing them to bind to each other, are said to be complementary. In particular, the nucleic acid bases that bind to one another stably are complementary.

complementary DNA (cDNA) A DNA molecule made from an RNA molecule by a special enzyme, reverse transcriptase.

complementation test A test to determine whether two distinct mutants are able to function normally when combined in a single cell, and thus to determine whether they affect the same gene or not.

conditional phenotype A phenotype that does not survive under one condition (restrictive condition) but is able to grow under a contrasting condition (permissive condition).

conjugation A process, especially in bacteria, during which two cells come into intimate contact and DNA passes from one into the other.

covalent bond A chemical bond formed by two atoms sharing one or more pairs of electrons, so considerable energy is needed to separate them, in contrast to some weak bonds that are easily broken.

crossing over The process in which two nucleic acid molecules (chromosomes) break and exchange segments.

cytosine One of the pyrimidine bases of DNA or RNA.

dalton A unit of molecular weight, essentially the mass of a hydrogen atom (specifically, one-twelfth the mass of a carbon atom).

deficiency A chromosomal mutation in which a segment of the chromosome is lost.

degenerate The genetic code is said to be degenerate because some amino acids are encoded by more than one codon.

deletion mapping A process of establishing a genetic map by using deletion mutants, based on the principle that if the extents of two mutations overlap they cannot recombine with each other.

denature To treat a molecule, such as a protein, in such a way that it loses its biological activity.

density gradient A solution (often of cesium chloride, CsCl) created in an ultracentrifuge tube or cell in which the density increases in the centrifugal direction, so that molecules may be separated by centrifugation on the basis of their size or density.

deoxyribose Technically, 2-deoxyribose; a variant of the five-carbon sugar ribose in which carbon atom 2 bears a hydrogen atom instead of the hydroxyl group, OH.

determination The process in embryonic development at which the fate of a cell is set; compare with *differentiation*.

differentiation The process in embryonic development in which a cell acquires its distinctive form and function; compare with *determination*.

dipeptide A molecule consisting of two amino acids joined by a peptide linkage.

diploid (adj.) Having two chromosome sets. (n.) An individual or cell that has two chromosome sets. Compare with *haploid.*

division pole In mitosis or meiosis, one of the centers at opposite sides of the dividing cell toward which the chromosomes move, and where a new nucleus will form.

DNA chip A small piece of glass with an array of single-stranded DNA molecules representing the elements of a genome, used to detect corresponding complementary RNA molecules.

DNA clone A vector (and cell in which it can replicate) carrying an inserted piece of DNA.

DNA ligase An enzyme that can form a phosphodiester linkage between two pieces of nucleic acid, thus joining them into one.

DNA fingerprint A picture showing the distinctive pattern of an individual's DNA fragments, made by cutting the DNA with a specific enzyme and separating the fragments in a gel.

DNA polymerase An enzyme that replicates DNA by linking nucleotides into a DNA strand.

dominant In the comparison of two alleles of a given gene, an allele whose effects are manifested when in the heterozygous condition; compare with *recessive*.

donor (1) In bacterial mating, the cell (Hfr or F^+) that donates DNA to a recipient (F^-). (2) In recombinant DNA work, the organism from which specific DNA is extracted for study or for insertion of genes into some other organism.

downstream On a nucleic acid molecule (chromosome), the direction of RNA transcription. Compare with *upstream*.

duplication A chromosomal mutation in which a segment of the chromosome is duplicated.

ectoderm In an animal embryo, the primitive tissue that will form the outer layers (skin) and much of the nervous system; compare with *endoderm, mesoderm.*

electrophoresis A process of separating molecules, such as nucleic acids, by moving them through a solid matrix (gel or paper) by passing an electric current through the matrix.

endoderm In an animal embryo, the primitive tissue that will form the lining of the intestine and related organs; compare with *ectoderm, mesoderm.*

endonuclease An enzyme that cuts a nucleic acid chain at an interior position, rather than removing nucleotides from one end.

endoplasmic reticulum A complex of membranes in a eucaryotic cell, especially parallel membranes that bear ribosomes and are a major site of protein synthesis.

end-product The compound, such as an amino acid, that a metabolic pathway produces.

enzyme A protein (or rarely, RNA) that acts as a biological catalyst by speeding up a chemical reaction in metabolism.

epigenesis A developmental process in which forms and structures are gradually constructed under the instructions of the genome.

episome A plasmid or viral genome that can exist independently inside a cell or in a state integrated into the cellular genome.

equatorial plate In mitosis or meiosis, the central plane along which chromosomes align during metaphase.

eugenics The ideal, and supposed science, of improving humans through selective breeding.

euphenics The process of ameliorating a diseased condition with a genetic basis by changing an affected person's phenotype, rather than changing the responsible genes directly. Compare with *eugenics.*

exon A portion of a split gene that does encode the sequence of a protein.

exponential growth A pattern of growth in which the ratio N_{g+1}/N_g is constant, where N_g is the number of organisms at generation g and N_{g+1} is the number one generation later.

express To actually use the information in a gene by synthesizing the encoded protein.

extinction In evolution, the process in which a population dies out.

F (fertility) factor A plasmid of *E. coli* that is able to promote genetic transfer from one cell to another.

F^+, F^- strain A strain of *E. coli* that either possesses (F^+) or does not possesses (F^-) an F factor.

fitness In ecological and evolutionary theory, a measure of the ability of organisms of a given genotype to reproduce themselves.

founder effect An effect in population migration in which the individuals who found a new population have an average genotype different from the average genotype of their parent population.

frameshift mutation A mutation in which the *reading frame* (*q.v.*) of a gene is shifted from normal.

functional genomics The science of determining how a genome is expressed.

gamete A sperm or egg cell, or their equivalents in more primitive organisms.

gap (G_1 or G_2) One of the periods in a eucaryotic cell cycle between the times of mitosis and DNA replication.

gastrula The stage in development of an animal embryo consisting of a hollow ball of cells in which cells from one side have pushed deeply into the interior to form internal layers.

gene knockout An experimental procedure in which a single gene is inactivated to determine its normal function.

gene therapy The procedure of correcting for a defective gene by introducing a normal gene into an organism.

genetic code The set of correspondences in which the sequence of nucleotides in nucleic acids specifies the sequence of amino acids in proteins.

genetic cross An experiment, or process, in which two types of organisms or viruses combine their genomes in a sexual or pseudosexual process, generally to determine the genetic characteristics of the resulting offspring.

genetic drift An evolutionary process in which the average genotype of a small population changes quickly, often in a non-selective direction, owing to random events.

genetic information The instructions encoded in an organism's genome by which it specifies the organisms's structure and operation.

genetically modified food (GM food) Food produced by a genetically modified organism.

genetically modified organism (GMO) A transgenic organism in which one or more genes have been introduced from an unrelated organism.

genome The total set of genes and regulatory signals that specifies the structure and operation of a virus or organism, encoded in a nucleic acid.

genomic library A set of DNA fragments from an organism, cut with a single endonuclease and spliced into vectors.

genotype A specification of the alleles of designated genes that an organism carries; compare with *phenotype*.

germ cell A cell from which the gametes develop.

germ line gene therapy A hypothetical, or proposed, cure for a genetic disease in which the disease alleles would be replaced in the germ (reproductive) cells, stopping their transmission to future generations.

glycogen A storage polysaccharide consisting of polymerized glucose molecules.

guanine One of the purine bases of DNA or RNA.

haploid (adj.) Having a single chromosome set. (n.) An individual or cell that has one chromosome set. Compare with *diploid*.

helix A figure with the form of the threads on a bolt, or of a spiral staircase.

heterozygote A diploid individual that carries two different alleles of a designated gene; compare with *homozygote*.

heterozygous Being a heterozygote.

Hfr strain A strain of *E. coli* bearing an *F factor* (*q.v.*) integrated into the cellular genome, so that it functions as a very efficient donor in conjugation.

homeotic gene A regulatory gene whose mutation produces a gross developmental change.

homolog Among the chromosomes in a diploid set, homologs (homologous chromosomes) are those that are of the same size and carry essentially the same set of genes.

homology A relationship among structures (anatomical or molecular) in different species that show similar or identical elements in similar or identical arrangements.

homozygote A diploid individual that carries identical alleles of a designated gene; compare with *heterozygote*.

homozygous Being a homozygote.

homunculus In the history of thought on reproduction and inheritance, a hypothetical minute person, contained in either the sperm or the egg, that was supposed to develop into a baby during pregnancy.

hydrogen bond A weak interaction capable of holding molecules together, in which a positive hydrogen ion is held between two relatively negative atoms, such as –O···H–O– or –N···H–O–, where the dotted line is the hydrogen bond.

hydrophilic "Water loving," a characteristic of a molecule that shows an affinity for water and tends to dissolve in water; compare with *hydrophobic.* See also *polar.*

hydrophobic "Water fearing," a characteristic of a molecule that shows an affinity for nonpolar substances, rather than water, and tends to not dissolve in water; compare with *hydrophilic.*

hydroxyl group The chemical combination OH.

inborn error of metabolism A genetic disorder caused by the lack of a critical enzyme, thus distorting a metabolic process.

induction (1) The process of expressing a gene by inactivating a repressor protein. (2) In embryonic development, the process in which cells of one type produce changes in cells of a different type.

information A measure of the orderliness or specificity of a structure.

interphase A term for the phase(s) of a eucaryotic cell cycle during which it is not going through mitosis.

intron An intervening, non-coding sequence inserted in a gene; compare with *exon.*

inversion A chromosomal mutation in which a segment of the chromosome is cut out and reinserted in reverse.

ion An atom or molecule that carries a positive or negative charge.

karyotype A picture of the chromosome set of an organism, made by spreading the chromosomes on a slide, photographing them, and aligning the homologous chromosomes.

kilobase A length of one thousand bases, nucleotides, or nucleotide pairs.

label (n.) A tag such as a radioactive atom added to some compound in order to follow its fate. (v.) To add such a tagged substance in an experiment.

lawn A layer of bacteria formed on an agar medium in which phage plaques may be seen; more generally, a similar layer of any cells used for visualizing virus plaques.

ligand A molecule that can bind specifically to a site on a protein.

linkage map A picture showing the chromosomal locations of genes and other genetic elements, determined by measuring their degree of linkage to one another.

linked Said of genes or other genetic elements that are on the same chromosome (nucleic acid).

lipid Roughly, a fat; a hydrophobic biomolecule, generally consisting of long hydrocarbon chains.

locus (loci) The position of a gene or regulatory element on a chromosome, or on a genetic map.

lysis The process in which a phage-infected cell (or more generally, a virus-infected cell) disintegrates and liberates the phage inside it.

lysogenic Adjective describing the condition of a cell in the state of lysogeny.

lysogeny A condition in which a bacterium harbors a temperate phage in a prophage state.

lytic cycle The process in which a bacteriophage multiplies in its host cell and lyses the cell.

macroevolution The large-scale changes observable in the fossil record, in which different types of organisms replace one another.

macromolecule A very large molecule, such as a typical protein or nucleic acid.

marker A mutation or other distinctive form of a nucleic acid, used to mark and define the position of a gene or other genetic element.

meiosis During a sexual cycle, the process in which a diploid cell divides its chromosomes into haploid sets.

membrane A structure made of protein and lipid molecules that surrounds and defines cells and various subcellular structures such as the nucleus and mitochondria.

mesoderm In an animal embryo, the primitive tissue that will form the bulk of internal organs; compare with *endoderm*, *ectoderm*.

messenger RNA (mRNA) An RNA molecule that carries the sequence of codons that specifies a nucleic acid.

metabolic pathway A series of chemical reactions, mediated by a series of enzymes, that transforms one substance into another step by step.

metabolism The sum total of all the chemical activities occurring in an organism.

metabolite One of the compounds being transformed through a metabolic pathway.

metaphase In mitosis or meiosis, the stage in which the chromosomes roughly align themselves in the middle of the dividing cell.

microevolution The small changes that occur within a single species.

mitochondrion A small, elongated body in a eucaryotic cell that contains much of the apparatus for extracting energy from food and storing the energy in a useable form.

mitosis In eucaryotic cells, the process in which the nucleus is precisely divided into two nuclei with identical sets of chromosomes, generally accompanied by cell division.

model An explanatory device in science consisting of a picture or description of the way some system might operate.

monomer One of the small molecules of which a polymer is made.

monosaccharide A single sugar, considered as a monomer of a polysaccharide.

mutagen A substance or phenomenon, such as irradiation, that increases a mutation rate.

mutant (n.) An individual that bears a mutation. (adj.) Bearing a mutation.

mutation (1) The process in which a change occurs in a genome. (2) The particular change resulting from such a process.

mutation rate A measure of the probability that a mutation will occur; generally, the probability per cell division that a given gene will mutate.

natural selection Differential reproduction of similar organisms with different genotypes, owing to their differential fitness for living in a particular ecological niche.

Neurospora A common red bread mold used for genetic experimentation.

nitrogen fixation A chemical process in which atmospheric nitrogen, N_2, is converted into a form such as ammonia, NH_3, or nitrate, NO_3^-, that can be used by plants.

nonsense mutant A mutant in which one of the nonsense or chain-terminating codons (UGA, UAG, or UAA) is created in place of a sense codon, thus terminating synthesis of the encoded polypeptide.

nuclear envelope A complex of two membranes that forms the boundary of the nucleus in a eucaryotic cell.

nucleotide A monomer of a nucleic acid, consisting of a sugar (ribose or deoxyribose) linked to a nitrogenous base and to a phosphate.

nucleus In a eucaryotic cell, the organelle that contains the chromosomes.

nutrient medium A mixture of substances that can support the growth of some organisms.

oocyte A cell that can undergo meiosis to produce an ovum (egg).

oogenesis The process of egg formation.

open reading frame (ORF) A candidate gene, identified as a series of codons between a typical initiating codon and a typical terminating codon, and large enough to encode a polypeptide.

operator A regulatory element adjacent to a gene or block of genes, forming the site where a regulatory protein can bind.

operon One or more genes whose expression is regulated by an operator.

ORF See *open reading frame.*

organelle In a cell, a distinctive structure with a specific function.

organic compound A compound based on one or more carbon atoms and commonly containing hydrogen, oxygen, and/or nitrogen; the type of compound of which organisms are made.

ovum An egg.

pangenesis The theory (now discredited) that sperm and egg cells are formed by taking substances from all parts of the body, thus accounting for the ability of these gametes to transport the characteristics of an organism to its progeny.

parental Having the same combination of alleles as one of the parents in a genetic cross; compare with *recombinant.*

pedigree A representation of a family (or its equivalent in other organisms) showing the pattern of inheritance of some characteristic(s).

peptide bond The linkage of C–O–N–H atoms that joins two amino acids together in a protein or peptide.

permissive The condition in which a conditional lethal mutant is able to grow, such as low temperature for a temperature-sensitive mutant; compare with *restrictive*.

phage A bacterial virus.

phenotype The expressed characteristics of an organism; compare with *genotype*.

phenylketonuria A genetic disease characterized by mental deficiency, owing to the inability to metabolize phenylalanine properly.

phosphate The chemical combination PO_4, found in nucleic acids and some other compounds.

phyletic evolution An evolutionary process in which a species gradually changes its characteristics.

phylogenetic tree A representation of the probable evolutionary relationships among a set of species, in the form of a branching tree.

plaque A small, circular clearing made by viral infection in a layer of cells growing on a nutrient medium.

plasmid A DNA molecule found in bacteria and some other microorganisms which can replicate itself, carries distinctive genes, can sometimes transfer copies of itself to another cell, but generally remains separate from the cellular genome.

plate To deposit organisms or viruses on a petri plate so they can be cultured and studied.

pleiotropic Of an allele that confers more than one identifiable phenotypic trait.

polar body One of the minute, non-functional products of unequal cell division during the meiosis of an oocyte (ovum).

polymer A molecule made by joining many similar or identical small molecules (monomers).

polymerase An enzyme that synthesizes a polymer.

polymerase chain reaction (PCR) A process of amplifying a small DNA molecule by repeated replication.

polynucleotide A nucleic acid; a molecule made by linking many nucleotides.

polypeptide A protein; a molecule consisting of many linked amino acids.

polyribosome A structure made of a messenger RNA molecule attached to several ribosomes.

polysaccharide A molecule made of many linked sugars (monosaccharides).

pool All the molecules of a given compound in the cytoplasm (or in a particular organelle) of a cell.

preformation The theory that an organism already exists in a minute but fully formed condition in a sperm or egg.

primary structure The sequence of amino acids in a protein chain.

probe A molecule, generally a small nucleic acid, that can be labeled and used to find a particular gene or other genetic element.

proflavin A type of dye whose molecules can bind to a DNA molecule so as to promote insertion or deletion mutations.

promoter A site on a DNA molecule (chromosome) where RNA polymerase molecules bind and initiate transcription.

prophage The form of a temperate phage genome, such as lambda or P1, when it is in the lysogenic state; commonly a circular form of the genome, integrated into the bacterial genome or free in the cytoplasm.

prophase The first phase of mitosis or meiosis, during which the chromosomes condense and the nuclear envelope disintegrates.

protein A polymer of amino acids.

prototroph An organism, such as a bacterium or fungus, that is capable of synthesizing its own components (amino acids, nucleotides, etc.) from simple carbon compounds such as sugars.

pseudovirion A virus particle that contains cellular DNA instead of the genome of the virus and so is capable of transduction.

pulse label An experimental regime in which a *label* (*q.v.*) is added to experimental cells or organisms for a short time, followed by a *chase* (*q.v.*).

Punnett square A square used to determine the results of a genetic cross by listing the gametes of one parent along one side, the gametes of the other parent along the other side, and combining their genes in each smaller square.

purine One of the nitrogenous bases, such as adenine or guanine, of a nucleic acid, characterized by a ring of six atoms linked to one of five atoms. Compare with *pyrimidine.*

pyrimidine One of the nitrogenous bases of a nucleic acid, such as cytosine, thymine, or uracil, characterized by a ring of six atoms. Compare with *purine.*

R plasmid A *plasmid* (*q.v.*) that carries genes that confer resistance to one or more antibiotics.

reading frame An imaginary way of marking off a triplet of bases along a nucleic acid to translate it into protein.

receptor A protein with a site specific for binding to some ligand whose function is to recognize the ligand and initiate some reaction to it.

recessive In the comparison of two alleles of a given gene, an allele whose effects are not manifest when in the heterozygous condition; compare with *dominant*.

recipient In bacterial conjugation, the F^- cell that mates with an Hfr or F^+ cell.

recognize To bind specifically to a certain molecule or a certain nucleotide sequence in a nucleic acid.

recombinant Having a combination of alleles that is different from either parent in a genetic cross; compare with *parental.*

recombinant DNA A combination made by cutting some DNA of interest with a restriction enzyme and combining it into a vector DNA cut with the same enzyme.

regulator gene A gene that encodes a regulatory protein.

repetitive DNA Any of the DNA, comprising a large part of typical eucaryotic chromosomes, that consists of short repeating sequences.

replica A copy that is essentially identical to the original, used especially with regard to nucleic acids.

replicate The process in which a nucleic acid molecule is converted into two essentially identical molecules.

repressor A regulatory protein that controls one or more genes by binding to a specific operator.

restriction endonuclease (enzyme) An enzyme that cuts DNA molecules in a specific way at a specific sequence.

restriction fragment length polymorphism (RFLP) A difference between individuals in the DNA sequences at some point in a chromosome, revealed by differences in the lengths of DNA fragments generated by a certain restriction enzyme.

restriction map A map made by ordering relative to one another the sites cut by various restriction enzymes.

restrictive The condition in which a conditional lethal mutant cannot grow, such as high temperature for a temperature-sensitive mutant; compare with *permissive*.

revertant An organism in which a mutation has been reversed by undoing the initial mutational change.

ribose A five-carbon sugar, the characteristic sugar of a nucleotide.

ribosomal RNA (rRNA) An RNA molecule that is a structural part of a ribosome.

ribosome A complex of protein and RNA molecules that serves as a cellular factory for protein synthesis.

RNA polymerase An enzyme that synthesizes an RNA molecule on a DNA template; the enzyme that performs transcription.

S period The period during the eucaryotic cell cycle during which the DNA of its chromosomes is replicated.

segregation The process in which the alleles of a gene separate into different gametes during reproduction.

semiconservative With respect to nucleic acid replication, a mode in which the strands of a double-stranded molecule separate and a new strand is synthesized on each original strand.

sibling A brother or sister.

sickle cell anemia A genetic disease characterized by red blood cells assuming bizarre sickle-like forms that clog small blood vessels.

sickle trait A mild form of sickle cell anemia characteristic of heterozygotes.

somatic cell In a multicellular organism, any cell (such as liver, muscle, or skin in an animal) other than a germ cell (a gamete or cell undergoing meiosis to form a gamete).

somatic gene therapy A process of correcting a genetic disease by introducing a normal gene into the tissue(s) affected by the gene.

speciation The process in which a species divides into two or more species.

spermatid One of the cells undergoing meiosis which will form spermatozoa.

spermatocyte A cell that is able to undergo meiosis to produce spermatozoa.

spermatogenesis The process of sperm formation.

spermatozoon (sperm) A mature male gamete.

spindle During mitosis or meiosis, a structure made of microtubules that separates the chromosomes to opposite poles of the cell.

starch A polysaccharide made of glucose molecules joined to one another in alpha 1:4 linkages, used as an energy storage material.

stem cell A cell in a multicellular organism that retains totipotency and the ability to produce cells that can differentiate into a variety of mature types.

stop codon A nonsense codon (UAG, UGA, or UAA) that signals termination of a polypeptide chain.

strain A specific type of organism or virus, generally of defined genotype, and maintained for research or breeding purposes.

structural genomics The science of analyzing the structure of a genome.

subspecies A distinctive and geographically limited part of a species.

substrate the compound that an enzyme acts upon; the compound that the enzyme transforms chemically.

suppressor A second mutation that reverses the phenotypic effect of an initial mutation.

synteny The condition of different organisms having similar or identical genes in the same position on their chromosomes.

telophase In mitosis or meiosis, the final stage, in which new nuclei form around the separated chromosome sets.

temperate Of a bacteriophage that is able to engage in lysogeny.

temperature sensitive Of a mutation or mutant that is functional at low temperature but not at a higher temperature.

template A form or molecule that can direct the formation of a complementary form or molecule.

template strand The DNA strand that serves as a template for RNA synthesis and therefore has a base sequence complementary to an RNA molecule.

test cross A cross designed to determine the genotype of an individual of dominant phenotype, by crossing the individual to a homozygous recessive individual; if the unknown is homozygous, only individuals of dominant phenotype will be produced, but if it is heterozygous some recessive offspring will be produced.

thymine One of the pyrimidine bases of DNA.

tissue In a multicellular organism, a structure composed of many cells of essentially the same type and with a distinctive function, such as muscle, skin, or a distinct nervous tissue.

totipotent In embryology, an adjective describing a cell that is capable of developing into any type of mature cell.

transcript An RNA molecule that has been formed as the complement of a DNA template.

transcription The act of forming a transcript, most commonly synthesizing a messenger RNA molecule carrying the information for a gene.

transduction The process in which a virus, generally a bacteriophage, carries cellular (genomic) DNA from one cell to another.

transfer RNA (tRNA) A small RNA molecule that carries an amino acid for protein synthesis.

transformation The process of changing a cell, generally a bacterium, by introducing DNA.

transgene A gene transferred into the genome of an organism unrelated to the source of the gene.

transgenic organism An organism that has been modified with a transgene.

translation The process of protein synthesis; compare with *transcription.*

translocation A chromosomal aberration in which part of one chromosome is attached to a non-homolog.

transposon A genetic element that can move from place to place in a genome.

turnover The process in which new molecules replace older ones; more generally, the process in which new objects of any type replace older ones.

ultracentrifuge A machine capable of spinning a solution or suspension at very high speeds, to create centrifugal forces powerful enough to move molecules through the solution.

upstream On a chromosome or nucleic acid, the direction opposite to the direction in which transcription occurs.

uracil One of the pyrimidine bases of RNA.

uridine A ribonucleotide containing uracil.

valence A description of the number of bonds an atom is capable of making with other atoms.

vector A virus, plasmid, or other structure capable of carrying a given piece of DNA.

vesicle A small membrane-bounded organelle that can carry material one from place to another in a cell.

virion The particulate, extracellular form of a virus, composed basically of a nucleic acid genome protected by a protein capsid.

virulent phage A bacteriophage that is only capable of multiplying in a lytic cycle.

virus A genetic entity, distinct from an organism and capable of reproducing only within a functioning cell, consisting of a genome and producing, at some stage, extracellular *virions* (*q.v.*).

wild-type In laboratory organisms, a standard, functional allele of a gene, in contrast to various mutant alleles.

X chromosome In mammals and some other animals, the chromosome that determines the female condition in the XX combination and is the pairing partner of a Y chromosome in males.

X-ray diffraction A method of determining the structure of a crystalline solid on the basis of the pattern it makes in scattering a beam of X-rays.

Y chromosome In mammals and some other animals, the male-determining chromosome that is the pairing partner of an X chromosome.

zygote The cell formed by the combination of gametes, generally a sperm and an egg.

notes

chapter one

1. Mark Schorer, *William Blake: The Politics of Vision* (New York: Vintage, 1959).
2. Joseph Haberer, "Politicalization in Science," *Science*, 178, 1972, pp. 713–23.
3. Theodore Roszak, *The Monster and the Titan: Science, Knowledge, and Gnosis*, *Daedalus*, Summer 1974, pp. 17–34.

chapter two

1. Robert Graves, "Introduction," in *New Larousse Encyclopedia of Mythology* (London: Hamlyn, 1959), p. vii.
2. Homer, *Iliad* V: 260–74, trans. E.V. Rieu (New York: Penguin, 1950), p. 99.
3. Virgil, *The Georgics*, trans. J.W. McKail (New York: Modern Library, 1950), p. 234.
4. Stuart Queen and Robert Habenstein, *The Family in Various Cultures.* (Philadelphia: J. B. Lippincott, 1974), pp. 158–9.

chapter three

1. We call an atomic weight unit a *dalton*, in honor of the British chemist John Dalton.

chapter four

1. We use the term *character* for a feature that may be inherited, such as eye or hair color, or flower color in a plant. A *trait* is a particular form of a character, such as purple flowers or white flowers.
2. Quoted in Sidney Schatkin, *Disputed Paternity Proceedings*, 2nd edn (New York: Matthew Bender, 1947), p. 199.

chapter five

1. Notice that both cells and chromosomes multiply by dividing: they double and then the two halves separate. Just as ships have always been called "she," it is standard in biology to say that the two duplicate pieces that are about to divide, or have just divided, are "sisters" and that they are "daughters" of the original single unit.
2. Kennedy McWhirter, "XYY Chromosome and Criminal Acts," *Science*, 164, 1969, p. 1117.
3. *Georgetown Law Journal*, "Note: The XYY Chromosome Defense," 57, 1969, pp. 892–922.
4. Barbara J. Culliton, "Patients' Rights: Harvard Is Site of Battle over X and Y Chromosomes," *Science*, 186, 1974, pp. 715–17.
5. H. Bentley Glass, "Endless Horizons or Golden Age?," *Science*, 171, 1971, pp. 28ff.
6. Ernest B. Hook, "Behavioral Implications of the Human XYY Genotype," *Science*, 179, 1973, pp. 139–50.

chapter six

1. Garrod himself realized the significance of the differences he detected but, as so often happens in science, others did not, and his work was not widely appreciated until several decades later when genetics had advanced far enough to make his work meaningful.
2. Robert Guthrie, "Mass Screening for Genetic Disease," in *Medical Genetics Today*, ed. D. Bergsma (Baltimore: Johns Hopkins Press, 1974).

chapter seven

1. Some strains of *E. coli*, notably one called O157: H7, have become notorious in recent years because they do cause severe – even fatal – intestinal diseases, but these are rare exceptions. The *E. coli* strains used in laboratories are harmless.
2. Felix d'Herelle, "The bacteriophage," *Science News*, 14, 1949, pp. 44ff.

chapter eight

1. Some of the phages that can grow on K are rII^{+} *revertants* that have experienced a back-mutation, but the rate of reversion can be measured independently and subtracted out; it is much smaller than the frequency of recombination.

chapter ten

1. Laurie Garrett, *The Coming Plague* (New York: Farrar, Straus & Giroux, 1994).
2. Bernard Dixon, "Antibiotics and Advertisers," *New Scientist*, 10 April 1975, p. 58.
3. Theodore Friedmann and Richard Roblin, "Gene Therapy for Human Diseases?", *Science*, 175, 1972, pp. 949–55.
4. Stanley Rogers, "Gene Therapy for Genetic Disease?," *Science*, 178, 1972, pp. 648–9.

chapter thirteen

1. Monsanto was one of the companies that launched a massive campaign against Rachel Carson after she published *Silent Spring*, which awoke the public to the environmental damage being done by pesticides and is generally credited with sparking the environmental movement.
2. This is not the place to debate economics, but those who place their faith in traditional economics should be aware that a growing school of economists is developing an *ecological economics* that repudiates many traditional ideas and places human economic activity in a wider biophysical context.

chapter fourteen

1. The strength of radiation is based on the amount of ionization produced in dry air and is measured in units of a *roentgen*. The biological unit is the *rem* (*r*oentgen *e*quivalent in *m*an), the dosage of ionizing radiation that will cause the same amount of damage as one roentgen of X- and gamma-radiation. A millirem (mrem) is a thousandth of a rem.

further reading

Brown, T. A. *Genomes.* New York: Wiley, 1999.

Cummings, M. R. *Human Heredity: Principles and Issues,* 4th edn. Belmont, CA: Wadsworth, 1997.

- A textbook on human genetics aimed at non-science majors.

Dawkins, R. *River Out of Eden.* London: Weidenfeld and Nicholson, 1995.

- An overview of the mechanism of evolution.

Griffiths, A. J. F., J. H. Miller, D. T. Suzuki, R. C. Lewontin, and W. M. Gelbart. *An Introduction to Genetic Analysis,* 7th edn. New York: W. H. Freeman, 2000.

- A general genetics textbook.

Guttman, B. S. *Biology.* Dubuque, IA: WCB/McGraw-Hill, 1999.

- A general introduction to the science of biology.

Klug, W. S., and M. R. Cummings. *Essentials of Genetics,* 3rd edn. Englewood Cliffs, NJ: Prentice Hall, 1999.

- An easy-to-read textbook on basic genetics.

Lewis, R. *Human genetics: Concepts and Applications,* 3rd edn. Boston: McGraw-Hill, 1999.

- A textbook on human genetics aimed at non-science majors.

Lewontin, R. C. *Human Diversity.* New York: Scientific American Press, 1995.

- A well-written, well-illustrated book aimed at the general public.

Sudbery, P. *Human Molecular Genetics.* Harlow UK: Addison Wesley, 1998.

- Describes some of the new technological approaches now used in human genetics, but is easy to understand.

Sykes, Bryan. *The Seven Daughters of Eve.* New York: W. W. Norton, 2001.
- A contemporary popular view of human evolution, based largely on DNA analysis.

Varmus, H., and R. A. Weinberg. *Genes and the Biology of Cancer.* New York: Scientific American Press, 1998.

Weiner, Jonathan. *The Beak of the Finch.* New York: Alfred A. Knopf, 1994.
- A fascinating story of contemporary studies of evolution.

Wills, Christopher. *Exons, Introns, and Talking Genes.* New York: Basic Books, 1991.
- An introduction to the work underlying the Human Genome Project.

index

Note: page numbers in *italics* refer to figures and tables